Postcolonialism, Indigeneity and Struggles for Food Sovereignty

This book explores connections between activist debates about food sovereignty and academic debates about alternative food networks. The ethnographic case studies demonstrate how divergent histories and geographies of people-in-place open up or close off possibilities for alternative/sovereign food spaces, illustrating the globally uneven and varied development of industrial capitalist food networks and of everyday forms of subversion and accommodation. How, for example, do relations between alternative food networks and mainstream industrial capitalist food networks differ in places with contrasting histories of land appropriation, trade, governance and consumer identities to those in Europe and non-indigenous spaces of New Zealand or the United States? How do indigenous populations negotiate between maintaining a sense of moral connectedness to their agri- and acqua-cultural landscapes and subverting, or indeed appropriating, industrial capitalist approaches to food? By delving into the histories, geographies and everyday worlds of (post)colonial peoples, the book shows how colonial power relations of the past and present create more opportunities for some alternative producer–consumer and state–market–civil society relations than others.

Marisa Wilson is Chancellor's Fellow in the Institute of Geography, School of GeoSciences at the University of Edinburgh, UK.

Routledge Research in New Postcolonialisms
Edited by Mark Jackson
Senior Lecturer in Postcolonial Geographies at the University of Bristol, UK

This series provides a forum for innovative, critical research into the changing contexts, emerging potentials, and contemporary challenges ongoing within post-colonial studies. Postcolonial studies across the social sciences and humanities are in a period of transition and innovation. From environmental and ecological politics, to the development of new theoretical and methodological frameworks in posthumanisms, ontology, and relational ethics, to decolonizing efforts against expanding imperialisms, enclosures, and global violences against people and place, postcolonial studies are never more relevant and, at the same time, challenged. This series draws into focus emerging transdisciplinary conversations that engage key debates about how new postcolonial landscapes and new empirical and conceptual terrains are changing the legacies, scope, and responsibilities of decolonizing critique.

Postcolonialism, Indigeneity and Struggles for Food Sovereignty
Alternative food networks in subaltern spaces
Edited by Marisa Wilson

Postcolonialism, Indigeneity and Struggles for Food Sovereignty

Alternative food networks in subaltern spaces

Edited by Marisa Wilson

Routledge
Taylor & Francis Group

LONDON AND NEW YORK

First published 2017 by Routledge

2 Park Square, Milton Park, Abingdon, Oxfordshire OX14 4RN
52 Vanderbilt Avenue, New York, NY 10017

Routledge is an imprint of the Taylor & Francis Group, an informa business

First issued in paperback 2020

British Library Cataloguing in Publication Data
A catalogue record for this book is available from the British Library

Library of Congress Cataloging in Publication Data
Names: Wilson, Marisa L. (Marisa Lauren), 1979– editor.
Title: Postcolonialism, indigeneity and struggles for food sovereignty: alternative food networks in the subaltern world / edited by Marisa Wilson.
Description: Milton Park, Abingdon, Oxon; New York, NY: Routledge, 2017. | Series: Routledge research in new postcolonialisms | Includes bibliographical references and index.
Identifiers: LCCN 2016020595| ISBN 9781138920873 (hardback) | ISBN 9781315686769 (e-book)
Subjects: LCSH: Food sovereignty—Cross-cultural studies. | Food supply—Social aspects—Cross-cultural studies. | Indigenous peoples—Food—Cross-cultural studies. | Postcolonialism—Cross-cultural studies.
Classification: LCC HD9000.5 .P64 2017 | DDC 338.1/9—dc23
LC record available at https://lccn.loc.gov/2016020595

ISBN: 978-1-138-92087-3 (hbk)
ISBN: 978-0-367-66809-9 (pbk)

Typeset in Times New Roman
by Keystroke, Neville Lodge, Tettenhall, Wolverhampton

Contents

Illustrations

Figures

Tables

Contributors

H M Ashraf Ali is Assistant Professor in the Department of Anthropology at the University of Chittagong, Bangladesh.

Melissa L. Caldwell is Professor of Anthropology at the University of California, Santa Cruz.

Stephen FitzHerbert is Senior Tutor in the Geography Programme at Massey University, Palmerston North, New Zealand.

Peter Jackson is Professor of Human Geography at the University of Sheffield.

Nicolette Larder is a Lecturer in the Division of Geography and Planning at the University of New England.

Amy K. McLennan is a Research Associate in the School of Anthropology and Museum Ethnography at the University of Oxford.

Charles R. Menzies is a Professor in the Department of Anthropology at the University of British Columbia.

Naomi Millner is a Lecturer in Human Geography in the School of Geographical Sciences at the University of Bristol.

Carolyn Morris is a Senior Lecturer in the Social Anthropology Programme at Massey University, Palmerston North, New Zealand.

Helen Vallianatos is Associate Professor in the Department of Anthropology at the University of Alberta, Canada.

Marisa Wilson is Chancellor's Fellow in the Institute of Geography, School of GeoSciences at the University of Edinburgh.

Sophia Woodman is Chancellor's Fellow in Sociology in the School of Social and Political Science at the University of Edinburgh.

Foreword

Moving beyond alternatives to recognizing multiplicity and complexity in food justice movements

Melissa L. Caldwell

In the coastal Northern California community where I teach and live, food sovereignty and alternative food systems are part and parcel of daily life. People from many different walks of life – students, farmers, activists, policy makers, scientists and ordinary consumers alike – come together in just as many diverse spaces – classrooms, municipal offices, supermarkets, community gardens, small family farms, large industrial vineyards and parking lots turned into farmers' markets – to debate and communicate their ideas about such pressing local and global issues as food access, fair labour practices, immigration, environmental sustainability and health, among many others. Food is more than simply a nutrient to be ingested, but is the vehicle through which community values and political and economic action become realized.

For many food activists like the ones in my own community, the neoliberalization of the global food system has become one of their most prominent concerns, and they actively promote the finding of alternatives that they believe will return American agriculture, nutrition and food cultures to more humane, equitable, safe and even traditional ways of life. School curricula for even the youngest American students now include learning how to garden and cook with 'natural' and 'organic' ingredients, and picking up a CSA (Community Supported Agriculture) box has become a regular part of weekly food shopping practices for many consumers. Spring break vacations for high school and university students have shifted away from the requisite beach trip towards a voluntourist experience in a foreign country – often in Central or South America – where students work alongside local farmers engaged in 'traditional' agricultural practices. The more paternalistic of these voluntourism enterprises invite budding student activists to share their experiences by helping local residents 're-learn' their lost farming and food processing traditions.

Largely overlooked in this embrace of alternative food systems and food sovereignty ideals is a careful interrogation of the concepts of either 'alternative' or 'sovereignty'. Too often, these concepts are simply taken for granted. Yet for people in many parts of the world, the 'alternatives' proposed by North American food activists – small, family farms; organic production techniques; private backyard gardens; heritage crops; and direct-sale or redistribution networks – are the mainstream modes by which they have produced, circulated and consumed foods for generations. And as such, it is not uncommon in places where these are already the standard way of doing things for ordinary people to want to find 'alternatives' as a way to express their own independence and self-knowledge.

In the case of Russia, where I have worked for many years, for friends and acquaintances who are exhausted after many years of doing their own gardening, making their own jams and preserves and haggling in farmers' markets where they have to navigate bloody and smelly fresh meat counters while wearing their good work clothes, the possibilities afforded by clean, quiet and orderly supermarkets are endless. Friends have confided how much more relaxing and appealing it is to engage in single-stop shopping, where they can trust the quality and safety of products that have been produced in technologically sophisticated and hygienically regulated manufacturing settings. Without the need to do their own gardening and processing, my friends enjoy having more personal time for themselves and their families. Just as importantly, the pro-liferation of grocery stores, restaurants and food products that was made possible by Russia's neoliberal capitalist transition represents the ultimate exercise of personal and national sovereignty: consumers now have access to choices and can make those choices on their own terms. For these postsocialist consumers, the neoliberal food system is a desirable alternative to a long cultural history of peasant-like farming, persistent shortages and state-controlled consumer practices (Caldwell 2009, 2014).

Such perspectives, however, do not always make sense to North American activists, scholars and enthusiasts. Several years ago I gave a lecture in which I discussed how state socialism shaped Russians' unique love/hate relationship with gardening and canning. By way of example, I related the story of a friend who was exuberant that, after her long hours as a physician, she no longer had to spend her evenings making baby food for her granddaughter but could simply go to the supermarket and buy jars of high-quality Western baby foods. Afterwards, a well-meaning food activist expressed his deep sorrow that my Russian friends had lost a sense of tradition. Had my Russian friend been there, I am quite sure that she would have expressed a very different opinion about the advantages of capitalist 'modernity' as an experience of freedom.

It is these different perspectives on the presumed nature and value of 'main-stream', 'alternative' and 'sovereignty' that are examined by the contributors to this volume. Curiously, this important intervention in the singularizing and moralistic approaches of most academic and activist food literature is long overdue. As the contributors show so convincingly, not only is it imperative to look for 'alternatives to the alternatives' (see also Jung *et al.* 2014), but also to understand how and why categories like 'mainstream' and 'alternative' have come to be and the political and cultural movements in which those categories are rooted. As Wilson, the editor of this book, writes in Chapter 7, 'Shared understandings of "mainstream" in relation to "alternative", and possibilities for collective moral and political action based on these shared understandings, depend in part on the ways people, places and things become embedded in global capitalist assemblages'. In other words, what the contributors to this volume persuasively demonstrate is that what counts as the norm of mainstream or its presumed 'alternative' antithesis is not an absolute given, but rather is the outcome of particular historical realities that are directly shaped by power dynamics on multiple scales and in multiple locations.

In the cases discussed here, these unique historical conditions and places that deserve consideration are the lived experiences of postcolonial and indigenous communities who do not fit neatly into prevailing preoccupations with the wealthy countries of the Global North – most notably, the United States, Great Britain and other 'Western' societies. What distinguishes this volume is a deliberate focus on how postcolonial and indigenous communities engage with the global industrial food system in ways that depart strikingly from one another, and often far differently than their Northern/Western counterparts. From South Asia and the South Pacific Islands to Africa and the Caribbean, the contributors critically illuminate how both local and global colonial and capitalist projects have shaped not just how local communities think about food production and consumption, but even what they count as healthy, safe and culturally appropriate food. The contributors raise numerous insights from their studies, but I will focus on several that I find especially productive and provocative for rethinking food justice concerns.

The first is that of knowledge politics. As each of the contributors shows, food sovereignty and alternative food networks projects are embedded within systems of power and inequality that privilege certain perspectives and voices, resulting in alienation, disenfranchisement and even exclusion. In Chapter 1 on Māori experiences with food sovereignty in New Zealand, Morris and FitzHerbert directly call attention to the problems encountered when outsiders speak for local communities, including when scholars engage in analyses of these issues. They show persuasively that conventional (read: élite, white, Western) ideas about the flaws of industrial capitalism and its presumed antidote of food sovereignty invoke an idealized 'indigenous imaginary' that is both the presumed possessor of more authentic and traditional modes of knowledge and the deserving victim in need of rescue. At the same time as these power dynamics turn social actors into victims, they also get upended when those social actors – in this case, Māori themselves – play their own 'indigenous imaginary' discourse to commoditize their own heritage products. At stake are not only questions about what constitutes authentic local knowledge and tradition, but also who is allowed to possess and use that heritage.

The simultaneous devaluation and revalorization of indigenous knowledge is a more universal problem, as shown by the case of El Salvador described by Millner in Chapter 4. In El Salvador, when non-governmental organizations and other development-oriented groups have attempted to promote local knowledge about agroecology and local crops, this has produced new versions of 'indigenous' culture that trouble identity categories and relationships between local and foreign. In the case of Bangladesh, Ali and Vallianatos in Chapter 2 critically show how long-standing traditions of colonial development approaches have led to a proliferation of forms of knowledge and knowledge brokers that in turn results in different actors – insiders and outsiders – having access to different audiences and sources of authority. Collectively these insights provoke bigger questions about how overlapping experiences of disenfranchisement and empowerment transect multiple registers and provoke uncomfortable questions about who is allowed to determine what counts as 'authentic,' 'traditional' and 'modern' and for whom.

Another important issue raised by these chapters is that of the effects of technocratic knowledge and practices, what James (2010) has termed 'bureaucraft'. Societies that have been on the receiving end of settler colonialism have also been subject to the imposition of external bureaucratic and technological systems of organization and regulation. In turn, these systems have directly shaped local forms of scientific knowledge, legal codes and property rights, among other areas, as well as pushing these local experiences into unexpected disputes at national and international levels. For the case of disputes over salmon fishing in British Columbia, Canada, Chapter 3 by Woodman and Menzies probes how indigenous communities have positioned themselves as defenders of 'salmon rights' in opposition to the Norwegian salmon fishing industry. At issue are disputes over which communities can make claims on ethical positions about sustaining the fish population versus sustaining human populations, and whether science-based environmental justice approaches trump cultural traditions. Chapter 5 by Larder shows how, across the world in Mali, West Africa, colonial bureaucratic systems introduced in order to stimulate agricultural production created a gendered division of agricultural labour. Whereas men were granted access to 'mainstream' cropping – in this case, irrigated rice production – women were relegated to 'alternative' crops (fruits, vegetables and other subsistence foods). As a result, not only were 'alternative' food systems marginalized and devalued, so, too, were the women associated with those systems.

This exclusionary nature of 'alternative' food networks and food sovereignty movements is especially intriguing, not least for prompting careful consideration of how Northern and Western proponents have touted the emancipatory and inclusionary potential of community gardens, organics, fair trade and their related initiatives. In every single chapter it can be seen that food sovereignty projects and alternative food networks divide communities as much as, and, in some cases, more than they bring people together in ways that are fair and equitable. Racial, class, gender and community differences are exacerbated, not lessened.

These divisive, exclusionary dynamics are perhaps most evident in Chapters 6 and 7 by McLennan and Wilson respectively. In Chapter 6, in an ethnographically vivid account of a post-mining community in the small Pacific Island nation of Nauru, McLennan explores why local residents prefer to purchase expensive, imported food products in supermarkets rather than work their family garden plots. What McLennan found was that gardens are contentious spaces. On the one hand, people rarely leave wills, which results in family disputes over land ownership, including gardens, which leads in turn to familial disputes over who bears the responsibility for working that land versus who is entitled to the products of that land. As a result, it is easier for family harmony when the land is left uncultivated. On the other hand, local residents are suspicious when foreign development project managers arrive to promote projects to work the land and hand out cash for those projects, and few are eager to accept resources from outsiders. Ultimately, buying expensive imports in a supermarket is the best option for maintaining family and community harmony.

In the case of Cuba and Trinidad, the subjects of Chapter 7, different histories of race and class have led to very different local appreciation for food movements. Trinidad's particular plantation history produced a unique system in which racial hierarchies among whites, free blacks, slaves and Indians played out through land ownership. Efforts to resolve these differences through struggles for racial and class unity have in turn shaped the food choices of local residents, both morally and materially. Most notably, processed foods are privileged as preferable to 'slave foods' grown locally. By contrast, Cuba's population has reflected different racial and class dynamics, especially vis-à-vis the country's shifting political ideologies from being aligned with the United States of America and, more recently, with the Communist Bloc. The communist or state socialist experience has facilitated ideals about the countryside and food access as a right and entitlement of citizenship. Wilson suggests that comparison of these two cases reveals important distinctions between thinking about whether consumption or production is privileged in civic-minded food projects.

Both individually and collectively, the chapters in this book provoke uncomfortable but important debates about the ways in which ostensibly 'just' food movements might, in fact, be far more complicated, divisive and destabilizing than they appear in the dominant Western imagination, whether scholarly, activist or popular. Yet the contributors do not seek to debunk and abandon food sovereignty, alternative food networks, or any of the many other food justice movements that exist in the world. Rather, their goal is to challenge rigid and reductive understandings of those ideologies by locating them in historical, political and economic context. In so doing, they expand our understanding of these systems and why they work (or not) in the places in which they are located.

This is an exciting collection, and one that will reposition the mainstream and the alternative in new configurations and dependencies. This is not so much a collection of 'alternatives to the alternative' as it is a conversation about multiplicity and complexity. With their critical insights and provocations, the contributors to this volume chart new directions for both scholarly and activist work. I hope that you find these conversations as generative as I have.

References

Caldwell, Melissa L. (ed.) (2009) *Food and Everyday Life in the Postsocialist World.* Bloomington, IN: Indiana University Press.

——(2014) 'Civil gardening in Russia: Growing a healthy Russia through ethical foods'. In Yuson Jung, Jakob A. Klein and Melissa L. Caldwell (eds) *Ethical Eating in the Postsocialist and Socialist World.* Berkeley, CA: University of California Press, 188–210.

James, Erica Caple (2010) *Democratic Insecurities: Violence, Trauma, and Intervention in Haiti*. Berkeley, CA: University of California Press.

Jung, Yuson, Jakob A. Klein and Melissa L. Caldwell (eds) (2014) *Ethical Eating in the Postsocialist and Socialist World*. Berkeley, CA: University of California Press.

Acknowledgements

The publication of this volume would not have been possible without the research participants who, in most cases, continue to live in their respective worlds in detachment from academic endeavours such as this. It is hoped that each of the chapter authors will share this work with their research participants, and that all of us (continue to) incorporate more inclusive ways of engaging with research participants so that they really become 'co-producers' of knowledge about sovereign foodways and other pressing issues of our time.

The editor would also like to thank a number of people who helped in brainstorming the subject matter in this book and in reviewing its chapters, including Peter Jackson, Melissa L. Caldwell, Isabelle Darmon, Amy McLennan, Isabel Fletcher, Valeria Skafida, Christine Knight, Stanley Ulijaszek, Laura Rival, Abigail Borron, Ben Garlick, Rebekah Miller and Deborah Menezes. The editor must also thank Faye Leerink (Associate Editor, Social Science Research, Routledge), Mark Jackson (Series Editor, 'Routledge Research in New Postcolonialisms' series) and the anonymous reviewers whose insights and suggestions have been crucial for sharpening the overall arguments in the book. These colleagues bear no responsibility for the views and interpretations of the authors and editor, however.

Introduction

Sovereign food spaces? Openings and closures

Marisa Wilson

The primary aim of this book is to explore connections between activist debates about food sovereignty and academic debates about alternative food networks (AFNs). AFNs are ideas and actions that in some way subvert or contest industrial capitalist foodways, such as urban farming, Community Supported Agriculture, agroecology, fair trade and so on, while continuing to work within its interstices. Similarly, food sovereignty emerged as a concept in activist circles (and only later in academia and policy) to describe the project of carving out separate or at least partially autonomous spaces for the production, exchange and consumption of food. In the context of this book, this means food projects based on a diversity of alternatives to industrial capitalist forms of accumulation and reproduction, representation and meaning, many of which 'transgress' into mainstream capitalist spaces (Goodman and Sage 2014). The chapters seek to capture this diversity and highlight the dynamic and contested *politics* of constructing alternative or counter-hegemonic spaces, a politics that is perpetually reshaped through everyday encounters including that between the researcher and his or her participants.

With notable exceptions,[1] the study of AFNs in academia has centred on knowledges, practices and state–market–civil society relations in the minority world (the global North) to the exclusion of a diversity of alternative–mainstream relations in the majority world (the global South). By contrast, work on and for food sovereignty has set the struggles of peasants, family farmers, fishermen and women, indigenous and landless peoples, pastoralists and rural workers in the majority world at the centre of global debates about non- or alter-capitalist ways of provisioning food.[2] Human geographers and other academics interested in alternative foodways have lagged behind in their efforts to uncover 'alternatives to the [Northern] alternatives' (cf. Goodman 2010: 115), and until now there have been no collective efforts to explore AFNs in postcolonial and indigenous worlds (though see related work on ethical foodways in the postsocialist world; Jung *et al.* 2014). By connecting work on AFNs with debates about food sovereignty, the volume seeks to develop a common ground from which diverse forms of engagement around these themes may emerge.

The volume also seeks to open up discussion about whether and how post-colonial and indigenous alternatives to capitalist food networks differ from each other and from those that have developed in high-income countries of the (former) imperial world, including food justice struggles of marginalized groups such as African Americans. Recent literature on food justice movements in the United States illustrates some of the ways that marginalized groups such as African Americans use food to enhance socio-economic wellbeing and empowerment, despite internal power relations and structural and ideological constraints (for example, Alkon and Mares 2012). While the latter face 'internal colonialisms' (Wald 1998; Elmer 2008) within the nation state that limit the scope for AFNs, postcolonial and indigenous peoples are emplaced differently in global and national economic and cultural orders, and so face other constraints and possibilities for the creation of alternative/sovereign foodways. Their contrasting histories and geographies of entry into global capitalist networks shape unique and diverse politics of alternatives, which emerge in the context of different 'alternatives' and 'mainstreams'.

The premise of the book is that, compared to many proponents of AFNs in the global North, postcolonial and indigenous peoples have distinct demands, values, beliefs and reasonings for contesting, subverting or indeed appropriating industrial capitalist foodways and different openings and closures for sovereignty and agency due to the *structural* and *epistemic* 'violences' of colonialism. In line with Farmer's (1996) usage, 'structural violence' refers to the ways in which histories and geographies of place make some people more susceptible than others to, in this case, negative socio-economic, health and environmental consequences of global industrial food networks. The associated idea of 'epistemic violence' (Spivak 1988) refers to the ways in which certain understandings and world views foreclose others, a process that occurs in the closed boardrooms of policymakers and development practitioners but also in scholarly work (a point I return to below). Both forms of violence (re)create what I call subaltern spaces: heterogeneous arenas of power that are nevertheless characterized by shared experiences of marginalization. By designating 'the subaltern' as a differentiated *space* rather than a homogenous group of people, the subtitle of this book stresses the spatially and socially dynamic power relations that override strict dichotomies between colonizer and colonized or dominant and resistant peoples, opening the way for potential alliances and 'creative appropriations' (Sivaramakrishnan 1995: 406) between differently positioned actors, including the researcher and researched. In this way, it is hoped that the book contributes to debates in Subaltern and Gramscian Studies about the unstable and processual nature of counter-hegemonic activity,[3] particularly how 'temporary unity' (ibid.) may be achieved through continuing struggles to articulate and activate a politics of sovereign food.

The great variety of openings and closures for food sovereignty in the postcolonial and indigenous worlds is also related to debates in human geography and other disciplines about 'diverse' economies and the diversity of AFNs.[4] In contrast to those who would discard the term 'alternative' altogether (for example, Goodman 2014) the authors continue to employ the term critically to enliven debates about the theoretical downsizing of capitalism (Gibson-Graham 1996, 2006), particularly in relation to subaltern struggles for food sovereignty. The authors are careful to distinguish the term 'alternative' from clear-cut acts of resistance, however, since some of the case studies (and related examples such as the Zapatistas, the Chipko movement in India and so on) are driven by material necessity as much as, if not more than, moral or political obligation (see Harvey 1996: 10, 187; Williams 2005: 8, 183; Abrahams 2007; Larder, Chapter 5). In this sense, distinctions between 'weak' and 'strong' forms of AFNs are useful in evaluating the different case studies of the volume: 'the former operate with an entrepreneurial culture that protects the environment [or contributes to local economies] whereas the latter highlight, among other aspects, the importance of labour conditions, local communities and small-scale farmers' (Gritzas and Kavoulakis 2015: 12). Of course, the relation between 'strong' and 'weak' alternatives is always relative, although one can certainly distinguish Cuban food sovereignty policies, which are grounded in 'strong' measures such as land redistribution and the official promotion of agroecology (Wilson, Chapter 7), from 'weak' AFNs such as indigenous potato production and marketing in New Zealand, which are more closely aligned to industrial capitalist forms of food provisioning (Morris and FitzHerbert, Chapter 1). The case studies in this book demonstrate how the divergent histories and geographies of people-in-place open up or close off possibilities for alternative/sovereign food spaces, illustrating the globally uneven and varied development of industrial capitalist food networks and of everyday forms of subversion and accommodation.

An overview of the chapters

With ethnographic examples from Africa, Latin America, the Caribbean, Asia and the Pacific, contributors to this volume seek to redress narrow understandings of 'resistance' to global capitalist networks by providing historically and empirically rich accounts of the ways indigenous (Chapters 1 to 4) and non-indigenous (Chapters 5 to 7) peoples of the (post)colonial world incorporate or reject mainstream foodways, under conditions that contrast with AFNs in the global North. In Chapter 1, Morris and FitzHerbert show how the Māori people of New Zealand utilize the discourses and practices of Pākehā (descendants of British colonial settlers) to (re)establish connections to indigenous ways of producing, exchanging and consuming food (in this case, potatoes). Similarly, in Chapter 2 Ali and Vallianatos argue that indigenous people in Bangladesh (the Pahari) carve out separate spaces for food production and consumption while incorporating some ideas and practices from foreign-led non-governmental organizations (NGOs), aid agencies and the state. As these authors illustrate, the Pahari people of Bangladesh continue to hold spiritual and environmental values for indigenous practices of *jhum* cultivation and food sharing, often in collaboration with economic development practitioners.

While the first two contributions to the volume highlight effective combinations of indigenous and market liberal agendas for food production and consumption, Chapter 3 outlines the de-colonial struggle of indigenous people in Canada to establish and institutionalize foodways that contrast fundamentally with western approaches. For indigenous people in Canada, salmon is more than just a productive asset or resource; it also symbolizes intimate and profound relationships that First Nation peoples have made over millennia with non-human actors. Woodman and Menzies argue that this understanding of salmon and its uses contrast with the kinds of Cartesian logics that separate environmental from economic concerns over salmon farming in places such as Scotland and Ireland. Similarly, Millner argues in Chapter 4 that indigenous peoples in El Salvador uphold views of the human and non-human world that contrast with output-driven agendas epitomized by the Green Revolution. Although it derives from external influences, permaculture offers an alternative model for agriculture that fosters hybrid knowledges and practices in El Salvador while enabling 'ecological autonomies' based on indigenized ways of knowing, doing and being.

Like the indigenous case studies, the non-indigenous chapters illustrate unique spaces of engagement and foreclosure for 'alternatives' that contrast with the North because of common histories of 'accumulation by dispossession' (Harvey 1982, 1996, 2003, 2005). Chapters 5, 6 and 7 critically investigate power-laden and differentiated processes of subversion, contestation and appropriation that stem from postcolonial histories and geographies of food production, exchange and consumption. Chapter 5 highlights struggles for 'everyday resistance', this time against a large-scale land grab in Mali that is, according to Larder, continuous with colonial projects to develop the Office du Niger into a rice-producing region. According to Larder, the emergence of AFNs in this region of Mali occurred, not through activist struggles, but because of the pre- and postcolonial exclusion of women from 'mainstream' rice production. Thus, while the women Larder interviewed developed what she calls 'alternative peasant trajectories', their practices of alternative food production are tied to a history of marginalization which contrasts with definitions of food sovereignty upheld by the Malian state and by Malian activists who advocate for a more radical version of food sovereignty.

The final two chapters highlight the *failure* of alternative foodways to emerge, McLennan for government-sponsored vegetable gardens in Nauru and Wilson for healthy, socially-just and sustainable foodways in Trinidad. As McLennan argues, consumers cannot develop the same kinds of relationships with producers on government-sponsored vegetable gardens in Nauru because the structural and symbolic logics that connect 'alternative' producers to consumers in places such as the USA and the UK have not developed in Nauru (or at least not in the same way). Instead, food networks in Nauru have long been broken by imported food and imported ideas, historical and geographical conditions that drive consumer preferences for unhealthy foods. Chapter 7 also speaks to food sovereignty debates about 'food literacy' – or preferences for healthy, sustainable foods – and the conditions under which they emerge, or do not (cf. Edelman *et al.* 2014: 917). As in other postcolonial 'developing' countries, food production in Trinidad remains industrial and export-oriented while food consumption is often tied to status hierarchies and ideas of modernity that lead to preferences for unhealthy food imports (see also Wilson 2013). These externally driven foodways are contrasted with those in Cuba, where small-scale food production has recently been associated with newer values of sustainability, consumption to longer-held values of shared subsistence (also see Wilson 2014a, 2014b).

Conditions of possibility

The ethnographic case studies in this book give contextual flesh to more general discussions about the uneven development of capitalism and its diverse temporalities and spatialities in the (post)colonial world. If capitalism universalizes, but is not imposed in the same way universally (Chibber 2013), then the subversion of industrial capitalism cannot be the same in places where this cultural economy first took root as a means to exact colonial tribute. In line with a relational (or dialogical) view of Self vis-à-vis Other (see p. 6, below), an implicit focus of the volume is on 'forms of resistance [that are only] plausible in advanced capitalist countries' (Spivak 1988: 288). Such forms of 'resistance' are partly tied to conditions of possibility brought about by colonialism. In Britain at least, many of the precursors of present-day AFNs (and other counter-hegemonic actions) would have been unthinkable without imperialist policies for production, trade and consumption. For instance, rural and guild reformers of early twentieth-century Britain were partly successful in their campaigns for state protections, craftsmanship and community farming because, in the context of the Great Depression, the British government shifted from a colonial *laissez-faire* food policy to a national 'Grow More Food' campaign. The latter was premised on the reinstatement of tariffs on food and other imports from the colonies (Blatt 2001) and colonial programmes for national self-sufficiency in places like Trinidad and Tobago (see Wilson, Chapter 7). Moreover, the successes of the British working-class movement – higher wages, the eight-hour workday and so on – were only politically feasible in Britain because, outside of the inter-war period, the country fed its workers with 'hunger foods' such as tea and sugar that derived from imperial trade networks (Mintz 1985). By appropriating land, labour and value from the colonies, Britain was able to establish what agricultural economist Alain de Janvry (1981) called 'sectoral and social articulation' between producers and consumers of products such as marmalade made from slave-produced sugar. In this 'national' economy, the production of capital and consumer goods in British factories was systematically connected to the consumption of working-class staples. Higher wages made good economic sense, since the primary consumers of marmalades, jams and jellies made in British factories were the working classes themselves.

In the colonial world, however, industrial food production and consumption were 'disarticulated' from their very beginnings. It did not matter whether the producers of raw materials were paid adequately, or at all, since their consumption was estimated as part of production costs and far removed from sites of factory production. Since only privileged groups had access to metropolitan goods, many inhabitants of former colonies came to associate these goods with modernity and progress. As McLennan and Wilson argue in Chapters 6 and 7 respectively, imperial trade relations continue to shape values for food imports in contemporary small island states of the Pacific and the Caribbean. The histories and geographies of imperial production, trade and consumption in these places established forms of structural and epistemic violence that continue to affect postcolonial peoples, arguably making them more susceptible to negative health and environmental consequences of the global industrial food system.

By delving into the histories, geographies and everyday worlds of postcolonial and indigenous peoples, this book seeks to show how colonial power relations of the past and present create more opportunities for some alternative producer–consumer and state–market–civil society relations than others. With this objective in mind, three related thematic strands emerge: (i) representing the 'Other'; (ii) articulating alterity; (iii) 'multiple sovereignties'.[5] In the rest of this introduction I will explain each of these in turn.

Representing the 'Other'

Rather than showcasing 'objective' data about subaltern peoples (Mizan 2014: 72), authors in this volume (who are mostly academics from the global North) focus on the politics of subaltern spaces that emerge through dialogical encounters and experiences, including those between the researcher and his or her research participants. By translating such encounters through their own logics, categories and meanings, the authors become part of the very politics of alternatives they write about, and so must be subject to critical reflections about how the 'Other' is or can be represented in such accounts. Following postcolonial and feminist scholars such as Spivak and Haraway, some of the authors (Morris and FitzHerbert, Millner, Woodman and Menzies), explicitly identify their own 'situated knowledges' (Haraway 1988) by outlining what their research and writing about the 'Other' can and cannot do. Though other contributors are less reflexive, as ethnographers, all the authors face the classic Maxwell's Demon problem identified by social anthropologists decades ago,[6] which places a double burden on those attempting to represent the 'Other'. The first is that both researcher and his or her participants are social beings who create meaning through verbal and non-verbal social categories, continuously restructuring these in everyday life according to changing events and circumstances, including the research encounter itself. This is why a dialogical view of this kind of research is so crucial. The second burden is that the ethnographic researcher seeks to understand, interpret and explain such processes of meaning-making and their everyday effects in the society under study through his or her own social categories, which are, in turn, products of the researcher's own social encounters and (often privileged) positioning in space and time.

As postcolonial and feminist scholars have long argued, representing the 'Other' is always a power-laden undertaking. However, this task has been charged to the contributors in order to initiate a dialogue about the conditions of possibility for alternatives which, in *all* places, are shaped by imperial and neo-imperial projects of capitalist accumulation and hegemonic cultures of modernity. Such a dialogue risks being one-sided if it is framed in such a way that it excludes other means and methods of engagement shaped by subaltern ways of knowing, doing and being. It is for this reason that emphasis must be placed on the *provisional* and *debatable* nature of food knowledges produced in this volume (cf. Jenkins 1994). Indeed, the authors and editor must be subject to a continuous 'state of answer-ability' (Noxolo *et al.* 2012: 425) that generates new lines of questioning to test the conclusions, leading to ongoing revisions and re-writes. This process must become more democratic than the current volume permits, so that our future research participants become active agents in the proposed dialogue (or another), developing ideas from their own frames of reference and in their own terms (ibid.). In Chapter 4 Millner intends to do just this through participatory perma-culture workshops with Salvadorian farmers. The next step would be to use these workshops to develop democratic forums for discussion such as blogs or community newsletters.

In its focus on dialogical relations between subaltern and dominant, alternative and mainstream meanings, values and practices, the case studies centre less on 'the subaltern's' inability to 'speak' (Spivak 1988) and more on whether and how certain powerful groups (state and market actors, global financial institutions, development practitioners, Western environmentalists, academics themselves) 'hear' messages (whether verbalized or not) about subaltern worlds:

> While many scholars in ethnic, indigenous and postcolonial studies have been troubled by the notion that marginalized subjects cannot 'speak', the focus on the inability to 'hear' opens up the possibility for building bridges across marginalized locations. Indeed, the salutary shift from the conditions of (failed subaltern) production to the conditions of (failed elite) reception is one of the things that makes the dialogue between postcolonial studies and indigenous studies simultaneously possible and desirable, as both movements struggle with how to articulate the tensions between overweening colonial power and resilient, resistant actors.
>
> (Byrd and Rothberg 2011: 5–6)

By largely focusing on those receiving messages rather than those sending them, chapters in this volume provide empirical depth to debates about 'representing the Other' in postcolonial, indigenous and Subaltern Studies. For instance, in Chapter 3, Woodman and Menzies argue that the alternative politics of non-industrial salmon farming upheld by scientists and environmentalists in Scotland and Ireland often exclude indigenous frames of reference. Similarly, they claim that scholarly accounts of industrial salmon production posed by actor network theorists fail to 'hear' indigenous perspectives that stray from the confines of the industrial network. Likewise, in Chapter 5 Larder argues that scholars of AFNs, who have principally focused upon examples from the global North, may misinterpret the kinds of 'alternatives' present in the Office du Niger in Mali. Although she poses a number of parallels between Northern and Malian AFNs – both, for example, centre on low-input production for local markets – Malian women's exclusion from land and productive inputs situates them in a subaltern space that is circumscribed by necessity rather than political conviction (also see Abrahams 2007).

This volume centres on how practices geared towards food sovereignty become expressed in various settings, with due caution about the power to name, compartmentalize and translate ethnographic material through the lens of one's own cultural and historical categories. Another theme of the volume, therefore, is how the politics of alternative/sovereign food becomes articulated, and particularly how this politics emerges through what Claude Lévi-Strauss called a 'bricolage' of Western and non-Western forms.

Articulating alterity

Postcolonial theorists such as Chakrabarty, Said, Hooks and Bhabha argue that knowledge and power, meanings and practices in the postcolonial world do not and cannot exist in a vacuum: they often draw from Western templates. 'Universalism is implicated in *both* imperial schemes to control the world and liberatory mobilizations for justice and empowerment. . . . Universals beckon to elite and excluded alike' (Tsing 2005: 9). Universal categories used to identify AFNs, such as 'organic', 'natural' and 'green', developed in Euro-American contexts through particular social understandings and material conditions. Among other influences, the twentieth-century organic food movement is tied to Austrian Rudolph Steiner's interest in Goethe's holism and Eastern philosophy, which shaped his school of biodynamic thought in the nineteenth century (Conford 2001: 40–70). Western tropes of 'nature' are embedded in the history of environmentalism and conservationism in the USA and Europe and their colonies (Peet *et al.* 2011). And the idea and value of all things 'green' cannot be separated from the histories of reactionary political movements in Europe (Bramwell 1989). In order to 'provincialize' such Euro-American social categories (Chakrabarty 2009), scholars must interrogate how articulations of alterity and their associated knowledges and practices enter into (and out of) diverse socio-political spaces.

While the circumstances under which AFNs developed in Euro-American contexts contrast with those of the (post)colonial world, the same universal discourses may be incorporated in both places. Thus Morris and FitzHerbert argue in Chapter 1 that the Māori of New Zealand use Western (Pākehā) discourses related to the 'authenticity' of Māori peoples to create a distinctive, marketable potato. In Chapter 2 Ali and Vallianatos illustrate effective combinations between indigenous Bangladeshi and market liberal discourses, which do not entirely undermine Pahari practices and meanings. And in Chapter 6 McLennan shows how local customs of land ownership and sharing on the Pacific island of Nauru clash with, but also emerge from, Western mandates for 'good' food production and consumption. Even the more clear-cut examples of alternative food networks in the book – Millner's Salvadorian permaculture in Chapter 4, Larder's 'alternative peasant trajectories' in Mali in Chapter 5 and Wilson's Cuba in Chapter 7 – are influenced by universal discourses and values that interact iteratively with local logics and practices, through what Anna Tsing (2005) calls 'friction'. This friction may foster the creation of new 'alternatives to the alternatives', as in

Millner's account of permaculturalists in El Salvador who appropriate but also change external meanings and practices to suit their own environmental and social context. As the term 'friction' suggests, however, cultural dialogue is played out in political arenas in which some (re)presentations are more powerful than others. In the case of Millner's El Salvador, voices of male brokers and those with political connections overpower other perspectives on permaculture. In the case of Wilson's Cuba, official designations of luxuries and needs define the nature and scale of sovereign food spaces and exclude other possibilities.

By addressing whether and how subaltern alternatives can be articulated and therefore politicized in practice, the volume seeks to highlight material and discursive processes that enable or disable projects for food sovereignty in subaltern spaces. Such a focus shifts attention away from Euro-American knowledges and towards wider debates about *global* experiences of self-determination and enclosure addressed by food sovereignty activists and scholars. It also points to multiple meanings of food sovereignty and its varied spatialities and temporalities.

Multiple sovereignties

This book seeks to illustrate how multiple sovereignties are performed through diverse, and sometimes competing, claims to self-determination. The primary difference between the two kinds of case studies in the volume (indigenous and non-indigenous) is the way in which sovereignty is imagined and actualized, particularly with regard to the most predominant 'sovereign' space: the modern nation state. In the non-indigenous case studies (Chapters 5, 6 and 7), there are definite structural and symbolic relationships between the construction (or not) of AFNs and the nation state. This is especially clear in both McLennan's and Wilson's accounts of state-led or sponsored producer–consumer networks that are more or less successful at drawing civil society to their cause. In the indigenous cases (Chapters 1, 2, 3 and 4), there is a more tenuous relationship between AFNs and national 'communities', which may be side-stepped through marketization (Chapter 1) or through the reproduction of indigenous epistemologies and ontologies that work separately from those implemented through institutions of state or market (Chapters 2, 3 and 4). In the latter cases, it is clear that hybridized forms that combine Western and postcolonial perspectives do not account for indigenous 'alternatives' as the latter are associated with completely different ways of being in the world.

By setting indigenous case studies alongside non-indigenous case studies, the book calls attention to the limitations of common definitions of 'sovereignty' which are set within the legal, political and moral parameters of the nation state. As anthropologists have long argued (Clastres 1990 [1974]; Dean and Levi 2003) the political structures of indigenous peoples evolved much earlier and very differently from the modern nation state. Legal, political and moral precepts such as rights, property and democracy, which are fundamental to modern nation states, look very different in indigenous contexts. While both indigenous and non-indigenous case studies in the volume reflect similar colonial histories of injustice, indigenous people are, in Kumar's (2011: 1568) terms, '*doubly* marginalized in postcolonial national cultures' because their struggles are based on the reclamation of resources from (former) imperial powers *and* nation states. Such peoples must struggle for a 'third form of sovereignty' (Bruyneel 2007, cited by Byrd and Rothberg 2011: 44) that contrasts with national or international strategies for liberal democratic incorporation and elides incorporation of the 'Other' through hybridization or bricolage.

In Chapter 3, indigenous people in Canada struggle to maintain traditional forms of salmon fishing in resistance to both national and international actors, some of whom use the discourse of 'rights' to neoliberal rather than collective ends. Multiple sovereignties are also evident in Chapter 2 which illustrates how indigenous Pahari groups are subordinated in their past and present relationships with Bengalis, who dominate state and market arenas. They are relevant to Millner's account in Chapter 4 of indigenous and *campesino* groups in El Salvador, who share histories of structural and epistemic violence vis-à-vis national and international actors. Although the other two indigenous case studies centre upon indigenous groups in the global North, each reflects similar spatialities of power. In Chapter 1 these are captured by contrasting interpretations of New Zealand's Treaty of Waitangi of 1840. While the Māori understood the treaty as a social exchange, the British version of the treaty's main principle – *kawanatanga* – saw the Māori ceding the majority of their land to the British colonial administration. Similarly, in Chapter 3, competing sovereignties determine how salmon as a resource is defined and allocated.

As struggles for food sovereignty take place in multiple spaces, so they are effected in different times. Though most postcolonial scholars emphasize the continuities of colonialism, indigenous and other scholars (for example, of the decolonial option) argue that the 'post' in 'postcolonial' is misplaced, particularly for indigenous peoples whose relationship to the nation state is, at best, complex.

> The uncompleted dialogue between postcolonial and indigenous perspectives is in part a result of the infamous and falsely periodizing 'post' in postcolonial: the misleading suggestion that colonialism is over. . . . Since confronting the ongoing colonization of native lands remains at the top of the agenda for indigenous peoples, many indigenous intellectuals have been reluctant to sign on to a theoretical project that appears to relegate their dilemmas to the past or an achieved 'after' (even if, in practice, this has rarely been the project of postcolonial studies).
>
> (Byrd and Rothberg 2011: 4)

Thus for indigenous peoples there is no single moment that marks 'before' and 'after' colonialism, such as national independence. It is for this reason that authors (such as myself) often place parentheses around the 'post' in (post)colonial, except when explicitly referring to the school of thought (postcolonial theory) or the time period after independence.

By highlighting the multiple sovereignties in which alternative food networks emerge, this book seeks to contribute to a growing rapprochement between post-colonial and indigenous scholars that centres on the need to recognize the various spatialities and temporalities in which colonial power is exercised. The chapters of the book reveal the multiplicity of forms of contestation and accommodation to foodways that emerge from the uneven development of agrarian and dietary space. Taken together, the chapters illustrate diverse configurations and politics of 'alternatives' vis-à-vis 'mainstreams' in subaltern spaces, where institutions, cultures and political economies of food, diet and agriculture differ markedly from places like the United States or the European Union.

As in most social scientific studies, however, this leads to more questions than answers. How, for example, do relations between AFNs and mainstream industrial capitalist food networks differ in places with histories of land appropriation, trade, governance and consumer identities that contrast with Europe and non-indigenous spaces of Canada or New Zealand? How do indigenous populations negotiate between maintaining a sense of moral connectedness to their agri- and acqua-cultural landscapes and subverting or indeed appropriating industrial capitalist approaches to food? This book explores these kinds of inquiry, in order to develop a more inclusive dialogue about the best ways to ensure healthy and sustainable food for all.

Notes

1 I.e., Whatmore *et al.* (2003); Goodman (2004); Hughes (2005); Ilbery and Maye (2005); Watts *et al.* (2005); Abrahams (2007); Maye *et al.* (2007); Caraher and Dowler (2007).
2 See, for example, *Via Campesina*'s ongoing campaign http://viacampesina.org/en/, the related Nyéléni Declaration of 2007 http://nyeleni.org/spip.php?article290 and recent special issues on food sovereignty published in (2014) *The Journal of Peasant Studies* 41(6), (2015) *Third World Quarterly* 36(3) and (2015) *Globalizations* 12(4).
3 I.e., Sivaramakrishnan (1995); Pandey (2011); Modonesi (2014).
4 I.e., Gibson-Graham (1996, 2006); Leyshon *et al.* (2003); Maye *et al.* (2007).
5 This term is used by Patel (2009), who is cited by Edelman *et al.* (2014: 918).
6 See Ardener (1989: 43–4).

References

Abrahams, C. (2007) 'Globally useful conceptions of alternative food networks in the developing south: the case of Johannesburg's urban food supply system'. In D. Maye, L. Holloway and M. Kneafsey (eds) *Alternative Food Geographies: Representation and Practice*. Bingley, UK: Emerald, 95–114.

Alkon, Alison Hope and Theresa Marie Mares (2012) 'Food sovereignty in US food movements: radical visions and neoliberal constraints'. *Agriculture and Human Values* 29(3): 347–59.

Ardener, Edwin (1989) *The Voice of Prophecy and Other Essays*. Edited by Malcolm Chapman. Oxford: Blackwell Publishing.

Blatt, Dan (2001) 'Descent into the depths (beginning in 1932): the collapse of World War I financial obligations', *Futurecasts* 3(7), 7 January www.futurecasts.com (accessed 11 November 2015).

Bramwell, Anna (1989) *Ecology of the Twentieth Century: A History.* New Haven, CT: Yale University Press.

Byrd, Jodi A. and Michael Rothberg (2011) 'Between subalternity and indigeneity'. *Interventions* 13(2): 1–12.

Caraher, M. and E. Dowler (2007) 'Food for poorer people: conventional and "alternative" transgressions?'. In D. Maye, L. Holloway and M. Kneafsey (eds) *Alternative Food Geographies: Representation and Practice.* Bingley, UK: Emerald, 227–46.

Chakrabarty, D. (2009) *Provincializing Europe: Postcolonial Thought and Historical Difference.* Princeton, NJ: Princeton University Press.

Chibber, Vivek (2013) *Postcolonial Theory and the Specter of Capital.* London and Brooklyn, NY: Verso.

Clastres, Pierre (1990 [1974]) *Society against the State. Essays in Political Anthropology.* New York: Zone Books.

Conford, Philip (2001) *The Origins of the Organic Movement.* London: Floris Books.

Dean, Bartholomew and Jerome M. Levi (eds) (2003) *At the Risk of Being Heard: Identity, Indigenous Rights, and Postcolonial States.* Ann Arbor: University of Michigan Press.

de Janvry, Alain (1981) *The Agrarian Question and Reformism in Latin America.* Baltimore, MD: Johns Hopkins University Press.

Edelman, Marc, Tony Weis, Amita Baviskar, Saturnino M. Borras Jr, Eric Holt-Giménez, Deniz Kandiyoi and Wendy Wolford (2014) 'Introduction: critical perspectives on food sovereignty'. *The Journal of Peasant Studies* 41(6): 911–31.

Elmer, Jonathan (ed.) (2008) *On Lingering and Being Last: Race and Sovereignty in the New World.* New York: Fordham University Press.

Farmer, Paul (1996) 'On suffering and structural violence: a view from below'. *Daedalus* 125(1): 261–83.

Gibson-Graham, J. K. (1996) *The End of Capitalism (As We Knew It): A Feminist Critique of Political Economy.* Minneapolis and London: University of Minnesota Press.

——(2006) *A Postcapitalist Politics.* Minneapolis: University of Minnesota Press.

Goodman, M. (2004) 'Reading fair trade: political ecological imaginary and the moral economy of Fair Trade foods'. *Political Geography* 23(7): 891–915.

——(2010) 'The mirror of consumption: celebritization, developmental consumption and the shifting cultural politics of fair trade'. *Geoforum* 41: 104–16.

——(2014) 'Eating powerful transgressions: (re)assessing the spaces and ethics of organic food in the UK'. In M. Goodman and C. Sage (eds) *Food Transgressions: Making Sense of Contemporary Food Politics.* Farnham, UK and Burlington, VT: Ashgate, 109–30.

Goodman, M. and C. Sage (eds) (2014) *Food Transgressions: Making Sense of Contemporary Food Politics.* Farnham, UK and Burlington, VT: Ashgate.

Gritzas, Giorgos and Karolos Iosif Kavoulakis (2015) 'Diverse economies and alternative spaces: an overview of approaches and practices'. *European and Urban Regional Studies*: 1–18.

Haraway, Donna (1988) 'Situated knowledges: the science question in feminism and the privilege of partial perspective'. *Feminist Studies* 14(3): 575–99.

Harvey, David (1982) *The Limits to Capital.* Oxford: Blackwell Publishing.

——(1996) *Justice, Nature and the Geography of Difference.* Malden and Oxford: Blackwell Publishing.

——(2003) *The New Imperialism.* Oxford: Oxford University Press.

——(2005) *A Brief History of Neoliberalism.* Oxford: Oxford University Press.

Hughes, A. (2005) 'Geographies of exchange and circulation: alternative trading spaces'. *Progress in Human Geography* 29(4): 496–504.

Ilbery, B. and D. Maye (2005) 'Alternative (shorter) food supply chains and specialist livestock products in the Scottish–English borders'. *Environment and Planning A* 37: 823–44.

Jenkins, Timothy (1994) 'Fieldwork and the perception of everyday life'. *Man,* NS 29(2): 433–55.

Jung, Y., J. Klein and M. Caldwell (2014) *Ethical Eating in the Socialist and Postsocialist World.* Berkeley and New York: University of California Press.

Kumar, Malreddy Pavan (2011) 'Another way of being human: "indigenous" alternative(s) to postcolonial humanism'. *Third World Quarterly* 32(9): 1557–72.

Leyshon, Andrew, Roger Lee and Colin C. Williams (eds) (2003) *Alternative Economic Spaces.* London, Thousand Oaks, CA and New Delhi: Sage Publications.

Maye, D., L. Holloway and M. Kneafsey (eds) (2007) *Alternative Food Geographies: Representation and Practice.* Bingley, UK: Emerald.

Mintz, Sidney (1985) *Sweetness and Power.* Boston: Penguin Books.

Mizan, Souzana (2014) 'Constructing the indigenous subaltern identity in the Brazilian media: ruptures in dominant representations'. *Polifonia Cuiabá* 21(29): 68–90.

Modonesi, Massimo (2014) *Subalternity, Antagonism, Autonomy. Constructing the Political Subject* (transl. Adriana V. Rendón Garrido and Philip Roberts). London: Pluto Press.

Noxolo, Pat, Parvati Raghuram and Clare Madge (2012) 'Unsettling responsibility: postcolonial interventions'. *Transactions of the Institute of British Geographers* 37(3): 418–29.

Pandey, Gyanendra (ed.) (2011) 'Introduction: the difference of subalternity'. In Gyanendra Pandey (ed.) *Subalternity and Difference: Investigations from the North and the South.* London and New York: Routledge, 1–19.

Peet, Richard, Paul Robbins and Michael Watts (2011) 'Global nature'. In R. Peet, P. Robbins and M. Watts (eds) *Global Political Ecology.* London: Routledge, 1–47.

Sivaramakrishnan, K. (1995) 'Situating the subaltern: history and anthropology in the subaltern studies project'. *Journal of Historical Sociology* 8(4): 395–429.

Spivak, Gayatri Chakravorty (1988) 'Can the subaltern speak?'. In C. Nelson and L. Grossberg (eds) *Marxism and the Interpretation of Culture.* Champaign: University of Illinois Press, 271–313.

Tsing, Anna Lowenhaupt (2005) *Friction: An Ethnography of Global Connection.* Princeton, NJ: Princeton University Press.

Wald, Patricia (1998) *Constituting Americans: Cultural Anxiety and Narrative Form.* Durham, NC: Duke University Press.

Watts, D. C. H., B. Ilbery and D. Maye (2005) 'Making re-connections in agro-food geography: alternative systems of food provision'. *Progress in Human Geography* 29: 22–40.

Whatmore, S., P. Stassart and H. Renting (2003) 'Guest editorial: what's alternative about alternative food networks?' *Environment and Planning A* 35: 389–91.

Williams, Colin (2005) *A Commodified World: Mapping the Limits of Capitalism.* London and New York: Zed Books.

Wilson, M. (2013) 'From colonial dependency to finger-lickin' values: food, commoditization and identity in Trinidad'. In Hanna Garth (ed.) *Food and Identity in the Caribbean*. London and New York: Bloomsbury, 107–19.

——(2014a) *Everyday Moral Economies: Food, Politics and Scale in Cuba*. Oxford: Wiley-Blackwell.

——(2014b) 'Agroecology and the Cuban nation'. In Yuson Jung, Jakob A. Klein and Melissa L. Caldwell (eds) *Ethical Eating in the Postsocialist and Socialist World*. Berkeley and New York: University of California Press.

1 Rethinking 'alternative'

Māori and food sovereignty in Aotearoa New Zealand

Carolyn Morris and Stephen FitzHerbert

The aim of this chapter is to consider what Māori struggles for food sovereignty in Aotearoa New Zealand reveal more generally about indigeneity and food sovereignty in a situation of postcoloniality. Through an exploration of Māori potatoes in contemporary Māori food sovereignty strategies, we question whether the term 'food sovereignty' necessarily goes with (or ought to go with) that of 'alternative', and raise the question of the politics that might be being performed by such an association. We show that many Māori do not necessarily make such a distinction, and argue that discourses and practices which reinscribe it may actually work against Māori sovereignty more broadly understood.

A core idea underpinning literatures on food sovereignty is that the capitalist food system will not deliver food sovereignty, if by sovereignty what is meant is the ability to have autonomy and control over the food you (as a group or individual) consume. Instead, sovereignty is imagined as emerging from spaces outside of the capitalist food system, spaces defined as 'alternative'. Efforts to bring food sovereignty into being have taken the form of assembling new (or reinstantiating imagined 'traditional') modes of production, distribution and consumption. Food sovereignty is considered to be manifest in assemblages such as farmers' markets, direct farmer/consumer linkages, gardening (individual and community), the Slow Food movement and so on. One of the sets of values that shape these assemblages are values of 'tradition', in which these practices are imagined as coming from (and therefore capable of reproducing) a better time, when food was wholesome, and people were connected rather than disconnected from each other and from the earth, and from knowledges of how to produce and cook food. A subset of this discourse of traditionalism is the indigenous imaginary.

The indigenous imaginary

In the indigenous imaginary (which has a very long and complex history) indigenous peoples are understood as being closer to nature than Western peoples. Being associated with nature rather than culture has, in the history of Western thought, generally been viewed as negative rather than positive as closer to nature has meant less developed, less civilized, less rational. However, in the field of alternative food, being closer to nature is valorized rather than denigrated. Being closer to nature means being in touch with what is authentic, original and uncontaminated (in particular by consumer capitalism). Constituted as the positive opposite of all that is wrong with the global capitalist food system, understood as possessing ecological and social wisdom and as having a spiritual connection with the land, indigenous peoples and their foodways are constituted as a reservoir of hope and knowledge from which an economically, socially, culturally and environmentally sustainable food system could be created. Bell (2014) draws attention to the pervasiveness of this imaginary in her discussion of the representation of the Na'vi in the film *Avatar*:

> The construction of Na'vi culture draws on long-standing stereotypes of indigenous that contrast their values and way of life sharply with those of capitalist modernity. The Na'vi live in harmony with nature, in contrast to the destructiveness of the humans' capitalist and technological engagement with the natural world. While the human society is driven by insatiable desires for more wealth, Na'vi society appears static, unchanging, maintaining balance with the natural world and connection with the spirits of their ancestors.
>
> (Bell 2014: 1–2)

The film *Avatar*, Bell writes, 'is testimony to the continuing power of the archetypes of noble, authentic indigeneity and rapacious modern, capitalist development' (Bell 2014: 2).

A number of writers (for example, Hage 1997; Abarca 2004; Morris 2010, 2013) have drawn attention to the problems associated with the positive discourse of authenticity and tradition for minority ethnic groups in white-dominated societies, noting the ways in which it works to contain and limit those groups, through constituting them (and the things associated with them) as objects for consumption by the dominant group rather than as economic subjects or actors. Bell writes that 'indigenous ways of life can only appear in modernity in the form of "tradition", appropriate for symbolic and ceremonial occasions, but not appropriate to the management of economic life, the organisation of social relationships, or the practice of government' (Bell 2014: 4). With regards to food this is not quite the case, as indigenous knowledges are imagined as a reservoir of desired, holistic wisdom, which can be drawn upon by the West to save itself and the planet.

In settler societies such as Australia, New Zealand and Canada the indigenous imaginary is complicated in interesting ways by the presence of actually existing indigenous people. Arguably, in societies lacking actual indigenous peoples, elements from the discourse of indigeneity (that is, traditional, natural, community) may be appropriated and deployed by food sovereignty activists to critique the existing capitalist food regime and to advocate a more sustainable future through the return to a more natural way of being and eating. In settler societies, however, the everyday presence of indigenous people complicates this process, as indigenous people repeatedly and consistently refuse to be contained by discourses of indigeneity which reflect the dominant groups' understanding of them, what they should *be* and what they should *do*. They refuse to be the dominant group's authentic 'other', however positively imagined, and instead insist that they can fully engage in the capitalist economy while also maintaining their traditional foodways. As Lambert writes:

Indigenous communities are not without agency in any economic space. Indigenous societies have opposed colonisation, attempted to subvert modernisation to their own purposes, and now – with the obvious hindsight of their previous experiences retained in social memory – continue to resist the marginal and passive status ascribed to them.

(Lambert 2008: 211)

Māori, like many indigenous peoples, insist on having their cake and eating it too (as well as making a business out of it and selling it, if they feel like it) and refuse to see this position as contradictory or inconsistent (see also Ali and Vallianatos, Chapter 2).

In this chapter we explore one aspect of how Māori, the indigenous people of Aotearoa New Zealand, are working to achieve food sovereignty, focusing on the case of the Māori potato. We argue that linking 'indigenous' with 'alternative' in opposition to 'mainstream' and 'capitalist market' reproduces a particular set of binaries (coded positively and negatively), and that this discourse is neither useful for understanding Māori practices nor politically useful for furthering Māori sovereignty.

Position, politics and methods

First, however, a note on our position. Neither of the authors of this paper are Māori, but rather Pākehā, descendents of British colonial settlers, and therefore part of the dominant group. There are significant challenges and responsibilities connected to the politics of knowledge production and the politics of represent-ation when studying Māori communities (Underhill-Sem and Lewis 2008; FitzHerbert and Lewis 2010); thus there is a complicated and contested politics to our work on potatoes, and writing 'about' Māori for our own ends (namely academic publication) is potentially problematic. As such, we keep to the surface and do not ask *matauranga* Māori questions, that is, questions refracted through Māori wisdom, or Māori ways of knowing; this is already being done by Māori writers (for example, Roskruge 1999, 2014; Lambert 2007, 2008; McFarlane 2007; Puketapu 2011). Our account is based upon two key sources of inform-ation: original ethnographic research by FitzHerbert and an interpretation of Māori voices in a variety of published literature about Māori potatoes. While none of this literature is directly concerned with concepts of food sovereignty, it speaks to the questions we wish to address, because of its focus on how Māori negotiate the articulation of diverse Māori economies and market economies (see, for example, Roskruge 1999; McFarlane 2007; Lambert 2008; Underhill-Sem and

Lewis 2008; FitzHerbert 2009, 2016; FitzHerbert and Lewis 2010; Bargh 2012; Barr and Reid 2014). In drawing on this literature to describe contemporary Māori engagement with potatoes, we problematize Western categories which constitute indigenous economies as separate from, and opposed to, capitalist economies.

Colonialism and postcolonialism in Aotearoa New Zealand

An understanding of contemporary Māori food sovereignty practices, and of the place of the potato in those politics, requires an understanding of the history of Aotearoa New Zealand and the place of the potato in that history. What this history shows is that 'Māori, like other indigenous societies, have never rejected innovation' (Lambert 2008: 204; see also Millner, Chapter 4), but instead have worked to achieve 'collaboration between two distinct knowledge bases: the globally fluid (if tensely negotiated) knowledge of horticultural production and marketing, and the locally fluid (and often equally tensely negotiated) practice of *matauranga* Māori [Māori ways of knowing]' (Lambert 2008: 201). What Māori history shows is that the binary that opposes culture and tradition to modernity and development does not capture Māori strategic agency in the face of challenges presented by ongoing colonial engagement.

Following the initial 'discovery' of Aotearoa and its naming as New Zealand by Dutch explorer Abel Tasman in 1642, and its subsequent rediscovery by English explorer Captain James Cook in 1769, New Zealand, and Māori, were incorporated into Britain's colonial empire. The Treaty of Waitangi, signed in 1840, formalized the relationship between Māori and the British Crown. Precisely what Māori signed up for and ceded remains a subject of intense debate, primarily because of the linguistic slippages between the English and Māori versions of the Treaty. In return for the protection of the Crown, Māori ceded *kawanatanga*, now translated as 'governance', but in the English version of the Treaty this word was translated as 'sovereignty'. Crucially, Māori did not cede *rangitiratanga*, which would have been a more correct translation of the English word sovereignty. The signing of the Treaty of Waitangi paved the way for mass migration from Britain, so that by 1858 Europeans outnumbered Māori. The resulting 'land grab' meant that by 1890 Māori, who had had undisputed possession of Aotearoa New Zealand, came to control only 40 per cent of the land, and that by the year 2000 Māori landholdings had been reduced to perhaps 4 per cent of the land mass. The loss of land that resulted from the Treaty of Waitangi 'strongly influenced Māori peoples' ability to have complete control over and direct self-determined, holistic development' (McFarlane 2007: 10; see also Durie 1998), resulting in the exclusion of Māori from the dominant economic sphere.

Colonization transformed the world of the Māori in many of the same ways it did everywhere, instituting 'political, technological and institutional changes that were overtly oppressive to Māori' (Lambert 2008: 211; see also Durie 1998; Smith 1999). With the incorporation of the Māori economy into the British capitalist economy, the economic base of Māori society and culture was rapidly transformed, with predictably negative consequences. As a result, Māori are now over-represented in many negative statistics: high rates of poor health, high rates of imprisonment, lower life expectancy, poor educational outcomes, low income and so on. However, this story is not the only story that can be told about Māori and colonialism. Māori were never passive in the face of colonial expansion. Despite their history, 'the alternative economic spaces of indigenous peoples remain, as evidenced by the endurance of their cultural economies' (Lambert 2008: 211; see also FitzHerbert 2009; McCormack 2013). From the beginning, Māori strategically worked to shape the colonial encounter to their own ends, seizing economic opportunities, indigenizing the new and resisting what they did not want. Resistance took many forms: non-violent protest, warfare, legal action, engagement with political processes and carving out spaces where Māori ways of being and doing continued to order the world.

The Treaty of Waitangi Act (1978) established the Waitangi Tribunal, whose mandate was to hear Māori claims of violations of the terms of the Treaty of Waitangi and to recommend settlements to the Crown. Initially the Tribunal could only hear claims for violations dating from 1978, but in 1987 the Act was amended, and the Tribunal was empowered to hear claims dating back to 1840. This Act marked a turning point in the Māori–Crown relationship. Māori seized the opportunity to seek redress for past wrongs and, since 2012, over one billion dollars has been paid out in compensation to *iwi* (tribes). As a result, those *iwi* have become significant economic actors, with the Māori economy now being valued at over forty billion dollars (BERL 2010). A key goal of *iwi* is to revitalize and sustain their society and culture, the achievement of which requires building a strong economic base. One strategy for achieving this aim is to mobilize certain things from the Māori economy into other economies (markets): for example, the Māori potato.

Potatoes in the history of Aotearoa New Zealand

In order to understand the place of the potato in contemporary Māori food sovereignty practices, it is necessary to explore the history of the Māori potato, because what is happening now is made possible (and limited) by what has happened in the past. Many Māori assert that potatoes pre-dated the European settlement of Aotearoa New Zealand (Harris and Niha 1999; Roskruge 1999; McFarlane 2007), but there is also widespread agreement that many of the potato varieties that have come to be known as Māori potatoes were introduced by early European explorers. It is not clear which explorer introduced the first potato – de Surville in 1769, Cook in 1769 or 1773 and du Fresne in 1772 have all been named – but one of the things that early European explorers in the Pacific did was introduce new animals (such as pigs, dogs and, inadvertently, rats), and crops (such as wheat, beans, carrots and potatoes),[1] to the places they visited (Lambert 2008). The key motivation behind this practice was to provide for the resupply of ships on future voyages (Lambert 2008), and the introduction of these new commodities opened up new economic possibilities for Māori.

The wide variety of potatoes labelled Māori potatoes (*Solanum spp*) in English are known generically by various Māori names in different parts of the country: *riwai, taewa, peruperu, parareka* (Harris and Niha 1999; McFarlane 2007). As Pākehā writers it seems most appropriate that we refer to them collectively as Māori potatoes.

While some sources show eighteen varieties of potatoes, other accounts list more. Linguist Bruce Biggs (1987), for example, lists fifty-three varietals, though he notes that some of these names may refer to the same sort.

Māori rapidly took up and integrated some of these new foods into their own foodways: 'It is remarkable how quickly certain exotic species were assimilated into Māori society, to the point where they were accepted without comment, i.e. were no longer innovations but almost mundane actors in the landscape' (Lambert 2008: 94). Prior to the introduction of the potato, the key starches in the Māori diet were the *kumara* (*ipomoea batatas*), a variety of sweet potato brought by Māori to Aotearoa when they arrived from Polynesia in the thirteenth century, and indigenous plants (for example, the *aruhe*, fernroot) (Roskruge 1999). By the 1830s potatoes had largely replaced these staples as they had several advantages: unlike *kumara* they could be grown throughout New Zealand, and they had higher yields and could be stored more easily (Harris and Niha 1999). Māori rapidly became skilled potato cultivators, as the cultivation practices for *kumara* could be applied to potato production, and by the eighteenth century they were reported as having potato gardens which exceeded fifty hectares, with some as large as 1,000 hectares. Potatoes provided Māori with new surpluses of food, which were used to feed an expanding population and to trade with other Māori and visitors to New Zealand. Charles Darwin, who visited New Zealand in 1835, noted that the potato was 'perhaps the greatest gift of the European to the Māori agriculturalist' (Roskruge 1999: 9). Indeed, while the origins of the New Zealand potato are debated, what is certain is that its adoption and utilization led to a Māori 'agricultural revolution' (Cameron 1964).

After the early explorers came European and American sealers and whalers to the Pacific, creating a demand for European-introduced crops. It is likely that these visitors introduced new potato cultivars, expanding the varieties available (Yen 1961, cited in Harris and Niha 1999). Māori were quick to engage in the growing opportunities for trade with Europeans:

> The stage was set for an entirely new form of Māori horticulture. Following the lead of several notable opinion leaders, the traditional economy (kin-centric, hunter/gatherer, horticultural) was augmented by new crop species as Māori were variously enrolling in an expanding market economy.
>
> (Lambert 2008: 93)

By the mid-1800s Māori had established a commercial horticultural industry, producing barley, wheat, maize and potatoes (Roskruge 1999; Petrie 2006): 'In 1857, the combined *iwi* of the Bay of Plenty and Tuwharetoa had upwards of 3,000 acres planted in wheat, 3,000 acres in potatoes, almost 2,000 acres in maize and upwards of 1,000 acres in *kumara*' (Firth 1929, cited in Roskruge 1999: 39). What is clear is that in the early days of the colonial encounter Māori were significant economic actors, who for a time were able to successfully operate within the expanding capitalist economy, largely on their own terms. However, demand by the burgeoning settler population for land led to conflict with Māori, resulting in the New Zealand Land wars in the 1860s and 1870s. Roskruge writes of the reduction in Māori participation in the primary sector, as they were 'diverted from their agricultural practices for a more urgent need, to support and assist their peers against the loss of their land' (Roskruge 1999: 41). The land confiscations that followed the wars – a loss to Māori of some 1,200 hectares of land – devastated their economic base and potential and, by the 1880s, Māori had returned to a subsistence lifestyle producing little more food than was necessary for their own consumption (Harris and Niha 1999; Roskruge 1999).

The revitalization of the Māori potato

While Māori were excluded from significant ownership in commercial horti-culture, gardening remained a significant activity (Leach 1984; see also Roskruge 1999), and it was in the domestic gardening sphere that the strains of the original potatoes survived. According to Lambert (2008), by the 1970s the label 'Māori' was applied to these non-commercial strains of potatoes, indicating that, while these varieties would also have been grown by Pākehā domestic gardeners, they had become associated with Māori. These potatoes were generally grown by *kuia* and *kaumatua* (elders) for *whanau* (family) or *marae* (ceremonial centres). A key value in Māori society is *manaakitanga* (hospitality), which is expressed through the provision of food at ceremonial gatherings at the *marae* (Harris and Niha 1999; Roskruge 1999; McFarlane 2007) and as such potatoes provided cultural as well as physical sustenance:

> For them the fact that they had 'Māori potatoes' was a thing of pride, especially as they were a *taonga* [treasure] which they felt had come from their *tupuna* [ancestors] and had been maintained by successive generations. To be able to produce these potatoes in the way they were taught by their parents and place them on the table for *manuhiri* [guests] and *whanau* was also a source of pride and added to their *manaakitanga* skills.
>
> (Roskruge 1999: 43–4)

Until recently, Māori potatoes were not often grown commercially by Māori, but there are now a small number of growers who cultivate on small to medium-sized farms to meet both cultural and commercial aspirations (see, for example, Lambert 2008; FitzHerbert 2009).

Māori potatoes in Māori development

As noted above, the settlements that came out of the Waitangi Tribunal claims process have provided a platform for contemporary Māori economic develop-ment. Māori are encouraged to, and desire to, develop previously undeveloped land

and resources and are also exhorted to, and aspire to, use this new wealth to address the social and economic problems they face as a group as a result of their history of colonization. The task for Māori, therefore, is to cultivate opportunities within (and for) their worlds (Durie 2003; Bargh 2007; Smith *et al.* 2015). One set of opportunities is provided by horticultural projects, including potatoes. The aim of such projects is self-determination, or sovereignty, and the tactic is twofold. The first is increased access to traditional food and food systems through the retention and utilization of traditional knowledge and practices and the production of healthy *kai* (food) 'as a vital part of *tikanga* [custom], culture and *whenua* [land]' (Moeke-Pickering *et al.* 2015: 38; see also Hutchings *et al.* 2012; McKerchar *et al.* 2015). The second is economic security through the generation of new income streams from Māori resources, and through the articulation of Māori and mainstream economies.[2]

The Māori potato is one example of this strategy (Roskruge 1999; McFarlane 2007; Lambert 2008; FitzHerbert 2009). The (re)emergence of Māori potatoes is largely due to the work of Tahuri Whenua (The National Māori Vegetable Growers Collective). This collective has generated a broader base of Māori potato growers, increased the supply of Māori potato seeds and spread knowledge about how to grow Māori potatoes (Roskruge 2004; Lambert 2008; FitzHerbert 2009, 2016). While not generally available in mainstream outlets such as supermarkets, Māori potatoes are increasingly on sale in alternative (gourmet and organic) food retail outlets. Stable commodity chains linking producers with retailers and consumers have yet to be assembled, meaning that availability is temporally and spatially sporadic and unpredictable. Additionally, a significant amount of the Māori potatoes found in such consumption spaces are grown by Pākehā producers. However, there is a potential market for Māori potatoes grown by Māori, as retailers we have spoken with report a demand for them. In the following sections contemporary Māori potato practices are described, showing that the potato has value both as a Māori object *and* as a Māori object to be exchanged, where value is located in both cultural and commercial worlds. It is argued that Māori do not necessarily see a contradiction in participating in the capitalist economy as indigenous people; commoditizing elements of their cultural heritage is not considered problematic for that heritage. What becomes clear is that the Māori engagement with Māori potatoes cannot be usefully understood through the alternative/mainstream capitalist binary. Illustrated below are multiple ways in which Māori use Māori potatoes, from forms of exchange that they (and outsiders) would regard as 'traditional' or 'cultural', to commerical production for mainstream markets. The cases described draw on FitzHerbert's 2009 research with four Māori potato growers. These accounts were generated from an ethnographic approach, where the author journeyed with Māori potatoes and growers, similar to Cook's (2004) notion of 'follow the thing'. Cases from the author's research are woven together with accounts from other key sources, namely the doctoral work of Simon Lambert, the Master's degree work of Turi McFarlane, and the research of Nick Roskruge, to provide a multi-faceted account of contemporary Māori engagement with Māori potatoes.

Contemporary Māori potato practices

Māori potatoes continue to be grown by Māori across Aotearoa (McFarlane 2007; Lambert 2008; FitzHerbert 2009; Puketapu 2011). The aspirations and practices of production and distribution differ across Māori garden projects and growers situate their Māori potato gardens within their own particular values and practices (McFarlane 2007; FitzHerbert 2009). Crop management practices vary between traditional Māori, organic and conventional, and often have some hybridity (McFarlane 2007; Puketapu 2011). There are a number of objectives at the heart of any Māori garden project, including: utilizing, and therefore preserving, Māori land, knowledge and practices; growing potatoes for *whanau*; growing potatoes for *marae*; growing potatoes to keep seed lines alive; and growing potatoes to sell (McFarlane 2007; Lambert 2008; FitzHerbert 2009). A theme shared across these documented accounts is that Māori values underpin potato production and that potatoes are grown to keep *Te Ao Māori* – the Māori world – alive. Indeed, growers consider the Māori potato a *taonga*. Interestingly, amongst the Māori elders whom FitzHerbert encountered, growers often mentioned it was not so important *how* Māori grew and exchanged their potatoes, as each grower applied their own particular cultural politics. Rather, they emphasized the importance of maintaining the overall practice of growing and exchanging potatoes for future generations of Māori (FitzHerbert 2009). As such, any garden project is always underpinned by more than simply commercial objectives, values and outputs; success is measured across multiple registers of cultural and economic value.

Table 1.1 illustrates contemporary Māori potato worlds in terms of the cultural goals, values and practices of Māori potato production and exchange as documented by Māori scholars (namely Harris, Roskruge, McFarlane, Lambert and Puketapu) and FitzHerbert. The physical reproduction of Māori potatoes is connected to cultural reproduction, embodying key Māori values in important cultural ceremonies. In terms of feeding the *marae*, *whanau* and *hapū* (kin group), growers gift potatoes for ceremonial and familial purposes. It is an act which provides cultural objects of significance for such occasions as well as offsetting the costs of hosting an event and feeding people. Indeed, growing and exchanging Māori potatoes is always about more than simply the provision of food. Instead, potatoes are generative of identity, value and relationships (Roskruge 1999; Harris and Niha 1999; McFarlane 2007). Roskruge (2009) notes that, for Māori, Māori potatoes are 'a part of who you are because you grow up with these foods. . . . They are part of what makes you unique'.

Table 1.1 provides a snapshot of the cultural world of Māori potatoes, although it is not an exhaustive (or homogenously representative) account. Its purpose is to introduce and make visible the multiplicity of Māori goals, values and practices which give Māori potato production its form and shape. What it shows is that potatoes embody and express Māori cultural values and economic values which emanate from outside of the Māori world, and that growers do not see this as contradictory.

Table 1.1 An overview of the cultural goals, values and practices attached to Māori potato production and exchange

	Contemporary Māori potato practices
Cultural goals	Retention of Māori land; retention of Māori knowledge; cultural sustenance; food sovereignty; self-determination; reinvigoration of Māori economy; utilization of Māori land; utilization of Māori vegetables; food supply; income generation; self-determination; establishment of Māori enterprise
Cultural values	Caring for and maintaining cultural heritage and objects (*taonga*); passing on heritage and objects to future generations; relationship to land and people; keeping a *taonga* alive; revitalization of cropping and exchange practices; pride in the knowledge potatoes live on after exchange
Cultural practices	Feeding *marae*, *whanau* and *hapū*; observing cultural practices and traditions; involving Māori youth in production; growing potatoes for *marae*, *whanau* and *hapū*; utilizing *whanau* labour; gifting seed to Māori schools and other Māori growers

Source: Adapted from FitzHerbert (2009).

Cultural and commercial values

Māori seek to (re)invigorate cultural assets and are increasingly looking to identify opportunities for utilizing Māori resources and generating new means of income. These opportunities involve mobilizing Māori resources such as the potato in new sets of economic circulations in which potatoes are commercially exchanged. As a number of authors have found, growers see the commercialization of Māori potatoes as an opportunity to utilize Māori land and inject investment back into that land (McFarlane 2007; FitzHerbert 2009). This is seen as a way to ensure economic survival, as well as the survival of Māori land and other cultural resources, such as seed lines and horticultural knowledges. McFarlane (2007: 32) shows that 'traditional horticulture provides a means for these growers to realize culturally-significant outcomes which provide unique opportunities for them to apply traditional Māori values and practices'.

Successful commercialization is based on the fact that the Māori potato offers something distinct from the broader potato market. Māori potatoes are visibly different from the majority of conventional potatoes. Their distinctiveness (in terms of colour and shape) is seen as a commercial opportunity, since they may appeal to emergent middle-class consumer values for traditional, authentic and traceable food (see Heldke 2003; Johnston and Baumann 2010; Morris 2010). As the Māori potato has become more widely known, for example through its use by New Zealand chefs as they work to create a distinctive New Zealand cuisine (see Morris 2010, 2013), it has emerged as something of a niche interest in the wider New Zealand foodscape. Māori growers see this as an opportunity; as one of McFarlane's informants stated, '[they] are probably worth more, . . . [and] have a higher value ... now you have people looking for these varieties and paying premium prices for them' (McFarlane 2007: 40). As Roskruge (1999: 67) suggests, Māori potato marketing could be 'aimed at the higher income consumer', and FitzHerbert has found that some Māori elders are amazed at the high prices which these potatoes command at places such as farmers' markets and speciality grocery stores (for example, NZ$8/US$5.2 per kilogram). Most growers have found that they have no problem selling their potatoes; indeed, they tend to be able to sell potatoes at a premium over conventional potatoes (McFarlane 2007; Lambert 2008; FitzHerbert 2009).

While there is a market for Māori potatoes, some Māori see a challenge in accessing this market and mobilizing market opportunities. The cost of market entry, the seasonality of Māori potatoes, the geographically dispersed localities of growers and the quantities produced, present limits to the desire and ability to commercialize production. McFarlane found that some growers believed that in order to realize the commercial potential of their potatoes, Māori would have to professionalize production:

> I don't think everyone can survive ... if they need to be productive and make a living out of what they know or what they are going to be taught then they have to be more professional in what they do. Like if you want to become a producer of these indigenous crops, you have to have a seller's market, you have to guarantee production and all these sorts of things. You can't just grow some in the back garden and hope you can supply a market somewhere without guaranteeing that you can supply a certain amount for a certain amount of time. You have to do things in a structured way, you can't just have a back-garden vegetable garden, you know, we have to become more professional, so that's where the young generations have to come in. .... We have to be professional if we want to make it economically viable.
>
> (McFarlane 2007: 40, citing informant)

As this citation illustrates, most growers experience no contradiction between growing *kai* for cultural purposes and selling commercially.

This ability to operate in (and for) multiple worlds can be seen in a number of growers' crop management practices. As illustrated by FitzHerbert (2009), Māori potato growers have successfully translated certain non-Māori ideas and practices into their potato enterprises, such as the use of herbicides, pesticides and companion planting. McFarlane (2007: 43) found that a number of growers considered the adoption of other practices as consistent with 'traditional Māori *tikanga*'. Lambert (2008: 222) argues that this integration is nothing new for Māori, and that 'rarely, if ever, was this framed in terms of being "non-traditional", or a threat to Māori culture'. From this perspective, the notion of a dichotomy between alternative and mainstream fails to capture how Māori engage in potato production and exchange.

It is important to note, however, that some Māori are criticized for combining approaches and motivations. One of McFarlane's grower-informants stated:

> You have always got those who don't want to mix it [Māori and Western knowledges]. It's definitely a minority and there is usually some ulterior story behind it. You know you get people that say you are a sell-out because you are selling our foods, and I'll say you can believe how you like, but the old people grew crops, so them to get by in this world [*sic.*]. What's the difference between us? The other thing is that if Māori don't take the bull by the horns then someone else will and then that opportunity is lost.
>
> (McFarlane 2007: 55)

What is of note here is that the Māori grower draws on 'tradition' to argue that commercialization is appropriate, and that he sees transformation and development as integral rather than antithetical to 'tradition'. Māori growers also consciously deploy 'tradition' in their efforts to create and control the Māori potato market. A number of Māori growers noted that the attributes of 'sustainable/ organic' and 'indigenous/native' were sought by the agri-food sector and had worked to mobilize this discourse for their own ends. As Lambert claims:

> For Māori, the very fact of their indigeneity enabled the simple differentiation of produce from other producers who were not Māori ... [indigeneity] enabled them to deploy an 'ethnic provenance' ... The symbolic lifeworld of Māori is the substructure upon which niche marketing and indigenous labelling, as purposive-rational actions within New Zealand's horticultural and cultural industries, are built.
>
> (Lambert 2008: 146, 189)

In order to take advantage of this symbolic opportunity, the idea of Māori branding has been mobilized by Māori collectives in relation to Māori tourism and Māori foods, including seafood, honey, dairy, kiwifruit and beef as well as potatoes (see, for example, Hutchings *et al.* 2012; Reid and Rout 2016). Lambert (2008: 214) notes 'the shadowy presence . . . of the "Noble Savage" discourse' in this project, with all the political dangers this presents, but what is key here is that this discourse is being deployed by Māori. This claim, and the practices described above, can be understood as assertions of Māori food sovereignty. Notably, sovereignty is considered by these growers to be found in both Māori and commerical market worlds.

Encounters with Māori potato growers

This section describes the exchange practices of four Māori potato growers, drawing upon FitzHerbert's research conducted in 2009. In the accounts, growers present their gardens and exchange practices during the 2008/2009 growing season. These four growers represent individual and collective enterprises and

Table 1.2 Taewa grower motivations and exchange institutions

Motivations	Non-formal exchange institutions	Formal market exchange institutions
• income • Māori employment opportunities • feeding *whanau* • utilizing Māori land • supplying *kai* to *marae* • inspiring other Māori • keeping *taewa* alive • serving traditional *kai* to visitors • teaching young people about gardening • developing a community garden • being less dependent on social welfare payments	• *koha* • gift • *marae* • *whanau* • *hapū* • community stores (trade in-kind) • farm gate sales (in-kind) • community gardens	• community stores (trade with money) • farmers' markets • farm gate sales (money) • supermarkets • internet sales • organic food stores (certified organic)

Source: Adapted from FitzHerbert (2009).

they all grow Māori potatoes for both *whanau* and *marae* and commercial exchange. The examples demonstrate how the exchange of Māori potatoes takes multiple forms for multiple purposes, and emphasize the many dimensions of economic practice, exchange and meaning. What becomes clear is that, individually and collectively, the modes of exchange in which growers work for self-determination transgress any alternative/mainstream binary (see Table 1.2).

The first of the growers FitzHerbert worked with, whom we will call Tahi,[3] grew potatoes on land she leased from a Pākehā. In previous seasons she had grown potatoes on *whanau* land (land collectively owned by her family) and *hapū* trust land (Māori corporately owned land). During the time of the interview, she continued to lease land from her *hapū* but also from a Pākehā. Her garden size had varied season by season, from a fifth of a hectare to two hectares. Tahi established the garden to feed her *whanau* and *hapū* but also to generate income for them by selling potatoes. She divided the potatoes she grew into three categories: potatoes for *whanau* (those of an irregular shape that are either too small or too large for commercial sale), potatoes for seed (those with the size and colour appropriate for growing next year's crop) and potatoes for sale (those of the right size and aesthetic quality for her markets). Tahi used the potatoes she kept for *whanau* to supplement household food supplies and to contribute to *whanau* and *hapū* gatherings at *marae* (for example, at funerals, family reunions or meetings). She often gave potatoes to elders who used them both for food and as seed to establish their own small potato gardens.

Tahi had established a number of market channels to sell her potatoes: a community store, which sells to people who live in the small settlement near her *whanau* land and to people travelling through the town, and a regional farmers' market held in a popular tourist town, which provided her with access to a broader base of customers. The farmers' market was a particularly profitable place of sale and Tahi regularly sold out of Māori potatoes. She noted that the people who go to the farmers' market (typically Pākehā and the middle class) had money and would pay a premium for Māori potatoes. The same NZ$8 per kilogram bags did not sell as well at the community store.

Tahi told FitzHerbert that her garden was by no means a 'hobby . . . I have to [sell potatoes]. It's a livelihood'. The sale of potatoes provided essential income (both individually and for *whanau*), and was used to pay the costs of leasing the land, the wages for *whanau* who work in the garden and also to purchase additional seed to supplement the next season's crop. As with the other growers, Tahi took great pride in the knowledge that her potatoes were desirable commodities in mainstream markets. She considered them to be an extension of herself, her land and her world – 'they are *out there* in the world, working for me, making connections'. Tahi thus exchanged potatoes across Māori and non-Māori institutions to meet her personal and cultural economic aspirations.

Rua, the second grower, cultivated two gardens, one for commercial production and the other for *marae* purposes. The first, a three-hectare commercial garden, was on her *whanau*'s privately owned land. While Rua's production on this garden was first contracted out for a major commercial vegetable processing company, she now produced Māori potatoes autonomously for supermarkets, distributors and seed suppliers. The enterprise was her main form of income and involved a partnership with a Pākehā who had 'expertise in the exchanging and marketing side of the business'. In order to maintain and increase her markets, Rua utilized the Horticulture New Zealand Good Agricultural Practice (GAP) certification standards. She was also involved in an emergent *iwi*-based commercialization project, which sought to create an internet-based market platform and Māori brand for Māori growers. Earnings from this garden were used to support Rua and her family.

Rua's second garden was one hectare and was a *marae* garden. It was located on Māori land adjacent to Rua's *marae* and was utilized to produce Māori potatoes for the family. The potatoes grown for the *marae* were exchanged through *koha*, a traditional gifting exchange system. *Koha* exchange involves transferring something in return for something considered to be of equal value; in this case, people exchanged money for Māori potatoes. This exchange may look like a commercial sale, but it is understood as a social exchange. There is no set price for the potatoes; rather, the person who wants the potatoes gives the amount of money they consider appropriate. Rua noted that, although the system generally worked well and that people gave a fair amount of money, there had been some instances when someone had provided no *koha* in exchange. However, she claimed that 'people pay what they can afford': a person may pay one dollar or five dollars since this exchange is a gift rather than a commercial exchange. The revenue generated from this garden was invested back into the garden. Surplus potatoes from both of Rua's gardens were used to provide food at cultural gatherings. Thus, like Tahi, Rua grew and exchanged Māori potatoes for both cultural and commercial purposes and used a variety of strategies to meet these goals.

The third grower, Toru, was a large-scale horticulturalist who produced potatoes on six hectares of *whanau* (family-owned land). The output from this garden, whether food or commercially generated money, was used to support *whanau* and *marae*. Toru had a number of strategies for maximizing his income. He used both New Zealand and Māori organic certification schemes, which support cultural practices and values and add commercial value to produce. He also spent considerable time investigating and negotiating market opportunities in supermarkets, restaurants, organic food co-operatives and food processing companies in order to provide flexibility and gain greater control by not being tied to any one contract. Another of Toru's strategies was to focus on value-added activities such as ready-made mixed salads, which included purple Māori potatoes. Essentially, Toru sought to identify a number of market channels that commanded the highest price for potatoes so that he could support his *hapū* by expanding their sustainable horticultural industry. Toru's aspiration was to utilize Māori land to generate employment opportunities for members of his *iwi*, to help 'draw Māori back from the city to their ancestral land'.

Whā, the fourth grower, farmed two hectares of potatoes on leased Māori land. Whā's emphasis was on commercial production, though any potatoes not suitable for commercial sale would be given to *whanau* and *hapū*. Similar to Toru, Whā spent a considerable amount of time investigating the potential of selling to the supermarket, where Māori potatoes commanded higher prices than conventional potatoes. He claimed that Māori-grown Māori potatoes produced added value since they were a premium product compared to non-Māori potatoes and Māori potatoes grown by non-Māori. As such he saw value in establishing a Māori brand for potatoes (and Māori products more generally), for which growers would produce potatoes according to their own localized customs and be able to tell their particular story on the packaging. This branding would distinguish the Māori-grown Māori potato and its packaging from others and enable shoppers to trace the potatoes back to their *iwi*. Whā aimed to sell Māori potatoes at supermarkets, fruit and vegetable stores and speciality organic stores.

While the four growers cultivated Māori potatoes to meet commercial aspirations, their fuller stories located these aspirations within the particular cultural context of each grower. For all growers, the cultivation of Māori potatoes is more-than-commercial – it is part of their practices and visions for keeping *Te Ao Māori* (the Māori world) alive. Motivations include sustaining Māori potato strains, providing for the food needs of *whanau* and *marae*, providing food for ceremonial occasions, selling potatoes to supplement income and generating work and income to support Māori self-determination, or sovereignty. Māori potatoes offer Māori growers something to grow and exchange based on the simultaneous value of the potato as a Māori object and as a Māori object to be exchanged. These growers mobilize multiple social and market institutions to serve their aspirations, working within and across multiple spheres of exchange, both capitalist and non-capitalist. These cases demonstrate the fluidity of the 'boundary' between Māori and mainstream economies, a fluidity that takes a particular shape according to the aspirations of individual Māori growers.

Conclusion: rethinking 'alternative' and indigeneity

The accounts shared in this chapter reveal the diverse practices and values which shape Māori engagement with Māori potatoes. They show that Māori practices of Māori potato growing cannot be made sense of through a set of categories that collapse indigeneity/alternative and oppose them to engagement with capitalist markets, since Māori 'supply both the newly established niche market for indigenous foods . . . *and* the cultural economy of small-scale traditional Māori institutions' (Lambert 2008: 131).

'Alternative' as a category is a convenient way of representing things not quite capitalist. However it often slips down a dangerous road of *re*-essentializing different ways of being and doing without attending to the various aspirations, negotiations and mobilizations of projects which escape being solely 'alternative'. What the case of Māori potatoes shows is both the relationship between formal and informal economies and the heterogeneity of 'alternatives' in Māori food economies (cf. Fuller *et al.* 2010). The case of Māori potatoes demonstrates how Māori growers reconcile and mobilize cultural and economic aspirations and cast Māori potatoes into circulations of both Māori and mainstream economies.

Unleashing Māori potatoes from certain binaries and oppositions offers researchers insight into the articulation of economies, and an opportunity to (re)-politicize the hopeful politics of alternative frameworks in ways more supportive of the initiatives of marginalized economic actors (FitzHerbert 2016). By contrast, constituting the mainstream, market-based capitalist food system as antithetical to political projects such as food sovereignty may limit what food sovereignty can be, by neglecting (and possibly disparaging) the practices of people who seek to achieve food sovereignty through engagement with mainstream markets. The premise that we live in either a capitalist or a non-capitalist system misses the complexity of how these systems are necessarily entangled or, more precisely for the case of Māori potatoes, how growers negotiate these worlds. In order to understand and support such negotiations 'we must attend to the ways that peoples' own struggles and visions of themselves create holes in dominant theories' (Biehl 2014: 116). If outside researchers such as ourselves are to contribute to, rather than limit, Māori aspirations, we must recognize and challenge our own theoretical categories and ways of knowing.

Notes

1   As with European potatoes, the varieties of potato now called Māori potatoes originated in Peru and Chile (Roskruge *et al.* 2010).

2 For example, dairying, forestry, fisheries, honey, tourism, and Māori certification schemes and brands (Reid and Principe 2008; Barr 2009; Hutchings *et al.* 2012; Barr and Reid 2014; Reid and Rout 2016).
3 All names are pseudonyms to protect the privacy of the growers.

References

Abarca, M. E. (2004) 'Authentic or not, it's original'. *Food and Foodways: Explorations in the History and Culture of Human Nourishment*, 12(1): 1–25.

Bargh, M. (ed.) (2007) *Resistance: An Indigenous Response to Neoliberalism*. Wellington, NZ: Huia Publishers.

——(2012) 'Rethinking and re-shaping indigenous economies: Māori geothermal energy enterprises'. *Journal of Enterprising Communities: People and Places in the Global Economy* 6(3): 271–83.

Barr, T. (2009) 'Indigenous authenticity, provenance and food branding'. Proceedings of the Agri-food Research Network XVI Conference, 24–6 November, Auckland, NZ.

Barr, T. and J. Reid (2014) 'Centralized decentralization for tribal business development'. *Journal of Enterprising Communities: People and Places in the Global Economy* 8(3): 217–32.

Bell, A. (2014) *Relating Indigenous and Settler Identities: Beyond Domination*. Basingstoke: Palgrave Macmillan.

BERL (2010) 'The Asset Base, Income, Expenditure and GDP of the 2010 Māori economy'. Report prepared for Te Puni Kokiri, The Ministry of Māori Affairs, Wellington, NZ.

Biehl, J. (2014) 'Ethnography in the Way of Theory'. In Michael Jackson, Arthur Kleinman and Bhrigupati Singh (eds) *The Ground Between: Anthropologists Engage Philosophy*. Durham, NC and London: Duke University Press, 94–118.

Biggs, B. (1987) *Complete English–Maori Dictionary*. Auckland: University of Auckland.

Cameron, R. (1964) 'Destruction of the indigenous forests for Māori agriculture during the nineteenth century'. *New Zealand Journal of Forestry* 9(2): 98–109.

Cook, I. (2004) 'Follow the thing: papaya'. *Antipode* 36(4): 642–64.

Durie, M. (1998) *Te Mana Te Kawanatanga – The Politics of Māori Self-Determination*. Auckland, NZ: Oxford University Press.

——(2003) *Ngā Kāhui Pou: Launching Māori Futures*. Wellington, NZ: Huia Publishers.

FitzHerbert, S. (2009) 'Following the Peruperu: geographies of circulation and exchange'. MSc thesis, The University of Auckland, Auckland, NZ.

——(2016) 'Geographies of economy-making: the articulation and circulation of *taewa* Māori in Aotearoa New Zealand'. Doctoral thesis, The University of Auckland, Auckland, NZ.

FitzHerbert, S. and Lewis, N. (2010) 'He Iwi Kotahi Tatou Trust: post-development practices in Moerewa, Northland'. *New Zealand Geographer* 66(2): 138–51.

Fuller, D., A. Jonas and R. Lee (eds) (2010) *Interrogating Alterity: Alternative Economic and Political Spaces*. Farnham, UK: Ashgate.

Hage, G. (1997) 'At home in the entrails of the West'. In H. Grace, G. Hage, L. Johnson, J. Langsworth and M. Symonds (eds) *Home/World: Space, Community and Marginality in Sydney's West*. Annandale, NSW: Pluto Press, 99–153.

Harris, G. and P. Niha (1999) 'Nga Riwari Māori – Māori Potatoes'. Working Paper No: 2/99. Lower Hutt, NZ: The Open Polytechnic of New Zealand.

Heldke, L. (2003) *Exotic Appetites: Ruminations of a Food Adventurer*. New York: Routledge.

Hutchings, J., P. Tipene, G. Carney, A. Greensill, P. Skelton and M. Baker (2012) 'Hua Parakore: an indigenous food sovereignty initiative and hallmark of excellence for food and product production'. *MAI Journal* 1(2): 131–45.

Johnston, J. and S. Baumann (2010) *Foodies: Democracy and Distinction in the Gourmet Foodscape*. New York: Routledge.

Lambert, S. (2007) 'The diffusion of sustainable technologies to Māori land: a case study of participation by Māori in agri-food networks'. *MAI Review*, 1, article 4.

——(2008) 'The expansion of sustainability through new economic space: Māori potatoes and cultural resilience'. Unpublished PhD thesis, Lincoln University, Lincoln, NZ.

Leach, H. (1984) *1000 Years of Gardening in New Zealand*. Wellington: A. W. & A. H. Reed.

McCormack, F. (2013) 'Commodities and gifts in New Zealand and Hawaiian fisheries'. In F. McCormack, and K. Barclay (eds) *Engaging with Capitalism: Cases from Oceania*. Bingley, UK: Emerald Group Publishing Ltd, 53–81.

McFarlane, T. (2007) 'The contribution of *taewa* (Māori potato) production to Māori sustainable development'. Unpublished Master of Applied Science in International Rural Development thesis, Lincoln University, Lincoln, NZ.

McKerchar, C., S. Bowers, C. Heta, L. Signal and L. Matoe (2015) 'Enhancing Māori food security using traditional kai'. *Global Health Promotion* 22(3): 15–24.

Moeke-Pickering, T., M. Heitia, S. Heitia, R. Karapu and S. Cote-Meek (2015) 'Understanding Māori food security and food sovereignty issues in Whakatane'. *MAI Journal* 4(1): 29–42.

Morris, C. (2010) 'The politics of palatability: on the absence of Māori restaurants'. *Food, Culture and Society* 13(1): 5–28.

——(2013) '*Kai* or Kiwi? Māori and 'Kiwi' cookbooks, and the struggle for the field of New Zealand cuisine'. *Journal of Sociology* 49(2–3): 210–23.

Petrie, H. (2006) *Chiefs of Industry: Māori Tribal Enterprise in Early Colonial New Zealand*. Auckland, NZ: Auckland University Press.

Puketapu, A. (2011) 'The lifecycle and epidemiology of the tomato/potato psyllid (*Bactericera cockerelli*) on three traditional Māori food sources'. Unpublished MSc thesis in Plant Protection, Massey University, Palmerston North, NZ.

Reid, J. and G. Prencipe (2008) 'The biological economy of Ngai Tahu Māori land-owners: hunters gatherers and scientists identifying pathways and challenges to building Nohoanaga Kaikai'. Proceedings of the 15th Agri-Food Research Network Conference, 26–8 November, Sydney, Australia.

Reid, J. and M. Rout (2016) 'Getting to know your food: the insights of indigenous thinking in food provenance'. *Agriculture and Human Values* 33(2): 427–38.

Roskruge, (1999) N. 'Taewa Māori: their management, social importance and commercial viability'. Unpublished Diploma in Māori Resource Development Research Report, Massey University, Palmerston North, NZ.

——(2004) 'He kai kei aku ringaringa'. Proceedings of the Te Ohu Whenua Hui a Tau, 8–9 July, Massey University, Palmerston North, NZ.

——(2009) 'Introducing taewa to New Zealand. Biotechnology Learning Hub' http://biotechlearn.org.nz/focus_stories/taewa_maori_potatoes/video_clips/introducing_taewa_to_new_zealand (accessed 11 November 2011).

——(2014) *Rauwaru, the Proverbial Garden: Nga-Weri, Māori Root Vegetables, their History and Tips on their Use*. Palmerston North, NZ: Tahuri Whenua.

Roskruge, N., A. Puketapu and T. McFarlane (2010) *Nga porearea me nga matemate o nga mara taewa: Pests and Diseases of taewa (Māori Potato) Crops*. Palmerston North, NZ: Tahuri Whenua.

Smith, G., R. Tinirau, A. Gillies and V. Warriner (2015) *He Mangōpare Amohia: Strategies for Māori Economic Development*. Whakatane, NZ: Te Whare Wananga o Awanuiarangi.

Smith, L. (1999) *Decolonising Methodologies: Research and Indigenous Peoples*. Dunedin, NZ: University of Otago Press.

Underhill-Sem, Y. and N. Lewis (2008) 'Asset mapping and whanau action research: "new" subjects negotiating the politics of knowledge in Te Rarawa'. *Asia Pacific Viewpoint* 49(3): 305–17.

Yen, D. (1961) 'The potato in early New Zealand'. *The Potato Journal* 1: 2–5.

2 Indigenous foodways in the Chittagong Hill Tracts of Bangladesh

An alternative-additional food network

H M Ashraf Ali and Helen Vallianatos

Introduction

In this chapter we explain how colonial histories and continuities in local, national and international development policies impact local economies and foodways in the Rangamati Hill District, Chittagong Hill Tracts (CHT) of Bangladesh. The expansion of capitalism was one of the central motivations of colonialism, resulting in the accumulation of economic capital by colonizers at the expense of colonial territories and peoples (Childs and Williams 1997; Loomba 1998; Harvey 2005, 2007). Since the Second World War, social and political movements against these processes of capitalist accumulation have led to an upsurge of new independent nation states. Nevertheless, the effects of colonialism on local and regional economies and social structures persist.

We examine the historical factors that shape contemporary food production, distribution and consumption practices among indigenous peoples (the Pahari peoples) in the Rangamati Hill District, CHT, situating these in political, economic and ethnocultural context. We argue that (post)colonial inequities are reproduced in contemporary food systems, but also that 'alternative-additional' (Fuller and Jonas 2003) food networks are created through such relationships. The notion 'alternative' is understood in terms of moments and institutions that exemplify Pahari resistance but also in terms of continued subsistence practices that centre on sharing food and other common resources. The notion of 'alternative-additional' is understood in terms of Pahari people's resistance to, but also partial adoption of, more 'mainstream', profit-oriented food practices. Sharing and exchanging food and other resources, as well as common land ownership, were central to Pahari economic and cultural norms before the British colonial administration introduced capitalist ideas of private property and the maximization of profit (Mohsin 1997; Nasreen and Togawa 2002). In contemporary Rangamati, these capitalistic motivations partly distinguish the food production and distribution systems of Bengalis and Pahari peoples, two ethnocultural groups who now co-exist in the CHT and plains of Bangladesh.

After providing a brief background to Rangamati as well as some ethnocultural differences between Pahari peoples and Bengalis, we briefly explain what we mean by postcolonial alterity in the context of development projects in the Rangamati region. We then examine how colonial and postcolonial economic and development policies shaped Pahari food networks by discouraging traditional Pahari farming practices such as *jhum* cultivation (shifting or slash-and-burn). Next we show how the more recent proliferation of a diverse array of non-governmental organizations (NGOs) has led to continuities of such policies and attitudes but also divergences and local resistance in the form of local Pahari NGOs owned by indigenous political leaders. In the final part we use ethnographic material to develop the argument that present-day Pahari food networks are becoming 'alternative-additional' as they reproduce traditional practices of food preparation and consumption and values of sharing but also incorporate new agro-technologies and profit-oriented relationships.

By explaining Pahari food networks in this way, we aim to contribute to a more inclusive understanding of variations in alternatives to the conventional global food system that respond to indigenous and postcolonial experiences and values. From this perspective, the notion 'postcolonialism' does not simply refer to a chronological event – the end of the colonial rule – but also recognizes colonialism and its effects as ongoing political, cultural and economic processes (Bhabha 1994; Childs and Williams 1997; Ashcroft *et al.* 1998).

For this chapter, ethnographic data were collected from different areas of Rangamati between May and August 2009 and February and July 2011. Methods also included semi-structured and focus group interviews with development practitioners and NGO representatives, which are drawn from for the relevant sections. All interviews were conducted in Bangla and transcribed and translated into English. Data analysis was done using ATLAS.ti.6.2, from which codes emerged that were sorted into categories and eventually into the themes that shape the chapter.

The Rangamati region and its ethnocultures

The Chittagong Hill Tracts (CHT) is the only mountainous region of Bangladesh. It was a single district until 1984 when it was divided into three separate districts: Bandarban, Khagrachari and Rangamati. Rangamati Hill District, where this research was conducted, is distinctive from the other two districts of the CHT because of the impacts of colonial and postcolonial economic and political policies on the local indigenous populations, some of whom were displaced by a hydroelectric dam. The CHT is now home to thirteen Pahari ethnic groups[1] and in our study we recruited research participants from a total of seven Pahari communities,[2] as well as Bengalis.

'Pahari' is not a single ethno-cultural group but a collective socio-political identity of all indigenous groups of the CHT (namely, Chakma, Marma, Tanchangya, Tripura and others, see endnote 1). It is not known who first used the term 'Pahari', but according to the narratives of some elderly leaders in the area, indigenous peoples in the hills coined it from the Bengali word *pahar*, 'hill', adding the suffix -*i* to indicate 'inhabitants of the hill'. Our research reveals a strong and ongoing attachment of Pahari peoples to their hilly environment. This sense of place is reflected in every aspect of Pahari economic systems: living patterns, housing types, food and agriculture practices, rituals, belief systems and gendered divisions of labour, which are markedly different from the ways of life of Bengalis in the plains (see Ali 2012). Pahari peoples claim that they are the original inhabitants of the CHT, as their ancestors lived in the hills long before the arrival of Bengalis. While many Bengalis are now living in the CHT, Pahari peoples are unwilling to recognize them as Pahari.

Relationships between Bengalis, the majority ethnic group in Bangladesh, and indigenous Pahari peoples, has often been antagonistic. After the Bangladeshi State's independence from Pakistan in 1971, Pahari peoples agitated for official recognition in the State constitution but were ignored by the Government of Bangladesh, who were all Bengalis. This was followed by the government's resettlement policies that moved Bengalis from the plains into traditional Pahari lands, with the support of the Bangladeshi army and with government subsidies for Bengalis. Pahari communal lands were not recognized by the government and were usurped in the resettlement process. The lack of political dialogue with Pahari peoples, despite their leadership efforts (see p. 41, below), resulted in some Pahari taking up arms. This resistance movement continued until the CHT Peace Accord was signed on 2 December 1997. After the Peace Accord, foreign NGOs implemented various development programmes for both Bengalis and Pahari peoples in Rangamati and other hill districts of the CHT (Mohsin 1997; UNDP 2009; Ali 2012).

Bengalis in the CHT have mobilized their knowledge of government policies and business practices, cultural affiliation with government bureaucrats and Bengali trade connections to dominate food trade in the area. For instance, to move many products out of the CHT, government permissions are required and Bengalis in the area are better able to work this system to their advantage. Bengali trade dominance contrasts with the lack of business culture and knowledge among Pahari peoples, who are used to communal ownership and the sharing of resources. Thus while Bengalis in Rangamati are employed in both formal (public service, service in NGO offices, and so on) and informal sectors (self-employment, daily labour and so on), Pahari peoples are primarily self-employed as farmers although some produce traditional textiles and a few are employed in the formal sector. It follows that, while the primary food producers in the area are Pahari, food markets are dominated by Bengalis.

Ethnic divisions are not only based on economic activities and values, but also on religious affiliations. While almost all Bengalis are Muslims, Pahari peoples are Buddhist, Hindu or Christian. These religious beliefs (particularly Buddhist and Hindu) are reflected in Pahari people's relationship to the land, the natural environment and the value placed on sharing resources and labour. Accordingly, Pahari farmers perform certain rituals before preparing land for *jhum* cultivation (see p. 44, below) or before harvesting crops. For example, Chakma villagers collect token money from each family in order to sacrifice an animal, such as a pig, goat or cow, for performing the Buddhist *thanamana* ritual for the 'god of place', so that their *jhum* land, crops and the villagers are protected from natural disaster and endemic diseases. Chakma farmers also practise *malayia dagana*, seeking help from others in their community for *jhum* cultivation and harvesting. *Malayia dagana* resembles what anthropologists call 'balanced reciprocity', which involves an exchange of goods or services between two individuals or groups. In this case, there is no need to make a formal agreement but individuals are morally responsible for providing similar services to persons or families from whom they have received services (labour, for example). This kind of exchange can be performed either immediately or during a certain period of time (such as during *jhum* cultivation or harvesting time). Chakma farmers believe that if a jungle is not cut, cleared and burnt on time a poor harvest (both in terms of quality and quantity) may result. So a farmer may seek his neighbour's cooperation to cut the jungle and to burn it and to prepare the land for *jhum* cultivation. This farmer may seek similar help from his neighbour for weeding *jhum* land and for harvesting crops. Farmers do not have to pay a wage for a neighbour's labour, but should offer a feast for those who help in performing these tasks or should do similar works for them if they need (as in *malayia dogana*). Pahari farmers in rural areas believe that this kind of practice enhances mutual cooperation and social solidarity among the villagers, which also pleases their god(s) and goddesses. Unlike the Bengalis, many Pahari farmers continue such social–cultural practices directly related to food production, distribution or sharing of resources in the rural villages of Rangamati.

The hilly topography of the CHT also contributes to differences in economic, social and cultural systems and lifestyles between Pahari peoples and Bengalis. As already mentioned, the Pahari have traditionally depended on a type of subsistence agriculture known as *jhum*, practised exclusively in the hilly areas. Traditionally, *jhum* agriculture maintained local ecosystems by managing communal forest resources, ensuring soil productivity and growing crops adapted to local microclimates. Because of the hilly landscape where *jhum* agriculture is practised, working small sections reduces soil erosion. When a section of jungle is cleared, vegetation is burned and the land is left fallow for a few months before planting. Sowing and weeding is done with minimal topsoil disturbance (Borggaard *et al.* 2003). A section of land is used for approximately two years, before being left fallow for between eight and ten years. While the fallow period allows the soil to rejuvenate, wild foods are collected from non-cleared areas. Thus all land is considered productive, providing different foods and sustaining local biodiversity.

In some areas of the CHT, the *jhum* agricultural system has recently changed in accordance with NGO and government development programmes geared towards commercial timber plantations and new agricultural technologies, such as power tillers, hybrid seeds, chemical fertilizers and pesticides (Gopal and Rasul 2005; Mallick and Rafi 2010; Ali 2012; Majumder *et al.* 2012). As explained below, the British colonial administration, postcolonial state and some contemporary NGOs see such modern agricultural technologies and the resulting commercialization of agriculture as more environmentally sustainable and economically efficient than *jhum* cultivation. By contrast, there is a common belief among Pahari farmers that traditional *jhum* cultivation can help regenerate vegetation and the soil and increase fresh food production and environmental sustainability. Below we focus on Pahari people's resistance *and* accommodation to such pressure to change their food production and marketing practices.

Postcolonial alterity

Colonialism was rooted in the social construction of indigenous cultures and peoples as primitive, uncivilized or inferior in relation to the colonizers (Nandy 1983). Research shows that the colonized accept some elements of the culture and ideology of the colonizers while rejecting others, in what Homi Bhabha (1994: 95) refers to as an 'ambivalent relationship'. Similarly, while the notion of 'alterity' implies that things or objects should necessarily be distinctive in relation to an 'other', its relationship with this other (for example, mainstream global economic systems) remains unavoidable, ambivalent, unequal and contradictory (Leyshon and Lee 2003: 196; Jonas 2010: 4). As with colonial ideologies, local people actively make and un-make existing economic systems in relation to changing situations and local needs.

In order to study the postcolonial alterity of food systems in CHT, we apply Fuller and Jonas' (2003: 57) distinction between three types of alternative economic spaces: '(i) alternative-oppositional, (ii) alternative-additional and (iii) alternative-substitute'. Alternative-oppositional economic spaces are based on a set of social values and ideals that actively resist hegemonic norms. Alternative-additional systems incorporate some elements of mainstream norms in their continuation of alternative or traditional economic systems (for analytical purposes, we treat the terms 'alternative' and 'traditional' as analogous). An alternative-substitute system may lose its distinctiveness, no longer preserving traditional institutional norms and values (cf. Fuller and Jonas 2003: 57). Our ethnographic and interview data suggest that Pahari food networks are alternative-additional. Thus while Pahari peoples continue to practise *jhum* cultivation and subsistence agriculture, a system that is not driven by market forces, some actors do adopt new agricultural technologies and participate in commercial food exchanges.

Building on ideas of postcolonialism and alterity, we show how both alternative/traditional and mainstream political economic foodways are reproduced and practised in the CHT in response to various economic and infrastructural development programmes and policies. Specific case studies, such as the impact of, and resistance to, the Kaptai Hydroelectric Project and Resettlement Policy, will be discussed and analysed in order to demonstrate how indigenous land rights, practices of *jhum* cultivation and values and ideas of sharing material resources are affected by such policies and programmes. We also examine how government agencies and microcredit NGOs encourage Pahari peoples to adopt capitalistic ideology of maximizing profit through the adoption of modern agricultural technology and the discontinuation of traditional *jhum* cultivation. We use such examples to show how Pahari peoples have become both 'traditional' and 'modern', preserving alternative/traditional food systems while adding modern technologies and elements of capitalist foodways as they see fit.

Colonial precedents

The CHT region was a British colony for almost two centuries (1760–1947) (Mohsin 1997).[3] In 1860, the entire CHT region came under direct colonial rule (prior to 1860, the British East Indian Company had regulated the CHT region).[4] In 1868 the doctrine of *terra nullius* created an entirely new forest management system consisting of reserve forests and unclassified State forests. Reserve forests denied (and continue to deny) Pahari peoples access for the collection of fuel wood, fodder and for the practice of *jhum* cultivation. Unclassified State forests were government-owned fallow land managed by the Ministry of Land and designated for meeting the agricultural livelihoods of the local population (see Rasul 2007). Like the postcolonial state, the British characterized the traditional farming system – *jhum* cultivation – as a primitive agricultural system that needed modernizing. Plough agriculture was the norm in Europe and it was assumed that this farming method was more productive and economically profitable, enabling the British to increase tribute collections from Pahari peoples (Schendel 1992; Nasreen and Togawa 2002; Tripura 2008).

The ongoing negative construction of Pahari peoples' *jhum* cultivation as backward (and sometimes unsustainable) has had ramifications for Pahari ideologies, values of resources and practices of communal landownership, particularly given the introduction of inter-ethnic relations which also occurred during the colonial period. In order to encourage Pahari peoples to adopt plough technology and cease the practice of traditional *jhum* cultivation, the British colonial administration brought Bengali farmers to the CHT region to teach Pahari peoples how to use plough technology. The British-backed Bengali farmers also brought new ideas of wealth accumulation such as the need to engage in surplus production and to reinvest in agricultural technology. Eventually, some Pahari communities, particularly Chakmas living in the Karnaphuli river valley region of Rangamati, adopted plough agriculture and began cultivating cereal crops, such as rice. However, plough technology did not affect those residing in the hills, especially the Tanchangya peoples. They continue *jhum* cultivation to this day, as explained by Manosh, a Pahari community leader:[5]

> Since ancient times our Tanchangya peoples have been producing foods through *jhum* cultivation. During the months of October and November they clean jungles in the hills and during the months of March and April they burn it. When the Baishakhi[6] storm clouds bring rains, our farmers begin to plant rice . . . and vegetables in *jhum* land with a chopper [a short ax with an iron blade]. During the months of May and June they will weed out the wild plants from the land to help grow the crops fast. By the end of July and August the farmers will harvest.
>
> (Interview on 2 March 2011)

Despite the persistence of *jhum* cultivation in some areas, colonial and postcolonial laws and policies affecting land use continued to cause dramatic economic and political changes, especially during the pre-1971 period when the country was still a part of Pakistan (see endnote 3). In 1953 and 1962, the Pakistani government undertook two large development projects: the Karnaphuli Paper Mills and the Kaptai Hydroelectric Project (also known as the Kaptai Dam), respectively. Both were financed by the World Bank and the United States Agency for International Development (USAID). The paper mills relied on raw materials such as bamboo and trees from deep in the forest and displaced thousands of Pahari peoples from their ancestral lands. The Kaptai Dam badly affected everyone living in the Karnaphuli River Valley and many other lowland areas of Rangamati. Forty per cent of the total arable lands in the CHT were submerged by the dam and more than 100,000 Pahari were displaced (Levene 1999; Nasreen and Togawa 2002).

Displacement disrupted subsistence practices and traditional ways of life. As one Pahari elderly participant, Badal Chakma, narrated:

> The fertility of our lands was very good and before the construction of Kaptai Dam we could produce food as much as we needed by investing minimal labour and capital. When crops were harvested we began to celebrate our various traditional festivals. It was through traditional practice that we entertained guests or neighbours with special items of food, or we shared our traditional food with each other in our village. We enjoyed our past time by dancing, singing folk songs and by telling stories to the children and young people about our ancestors. Our Chakma minstrel poet described our ancestors, how they originated, where they were living and how they came here through singing songs. But when the Kaptai Dam was constructed in the Karnaphuli river we lost almost everything: our land, our traditional culture, our food and economy, and we became refugees.
>
> (Interview on 30 July 2009)

Following the war of independence from Pakistan in 1971, the new Government of Bangladesh continued policies implemented by prior administrations. The Resettlement Policy of 1979 legalized the resettlement of Bengali landless peoples from the plains to the CHT. Many of these Bengalis were also displaced by development projects such as the construction of the Jamuna Bridge near Dhaka.[7] Like prior governments, the Bengali government assumed that land in the CHT was available in accordance with the idea of *terra nullius*, ignoring traditional Pahari communal ownership and usurping Pahari lands for Bengalis. It is possible that shared histories of displacement among the resettled Bengalis and Pahari peoples fostered a willingness to work together, although racial tensions still exist between the two groups. By the mid-1980s it is estimated that more than 400,000 Bengalis had been relocated to traditional Pahari territory (UNDP 2009). One research participant explained how Bengali displacement and migration was understood by the Pahari:

There is a Chakma proverb – '*lorne morne saang*' [changes in ancestral lands may lead to suffering or death]. This is very applicable to those who had been displaced by the Kaptai Dam or because of the Resettlement Program . . . The resettlement policy . . . created huge social and economic problems for the Paharis as the Bengalis started to settle in their [Paharis'] lands with the support of Bangladesh army. Then these Bengalis began to encroach the *khas* lands and forest resources, which were previously under the common ownership of the Paharis. Government people told the local Paharis in Rangamati that all the [remaining] cultivable lands [that were not entirely submerged by] the Kaptai Dam would be returned to the actual owners but the government gave these lands to the Bengalis settlers.

(Interview on 28 June 2009)

Postcolonial nation states of Pakistan and Bangladesh continued to prevent Pahari peoples from accessing forests for *jhum* cultivation and for collecting forest resources such as wood for fuel. By the 1980s, the amount of Pahari lands and resources had become extremely limited due to displacement by the Kaptai Dam, the resettlement programme, reclamation of land by the government for their own needs (such as army camps, government buildings, tourism resorts) and population growth. The result was a reduction in local biodiversity and a change in the economic life of local Pahari peoples. As the British colonial administration had done, so the Government of Bangladesh and its development agencies placed blame for the decreasing productivity of lands on Paharis' *jhum* cultivation.

Resistance and accommodation: NGO development programmes and changing food production in the CHT

Pahari political leaders resented the Government of Bangladesh and Bengalis because of the political decisions and development programmes described in the previous section. This resentment culminated in armed resistance in 1975, resulting in minimal access to the CHT for outsiders because special government permission was required for entry into the area. The indigenous resistance movement continued from 1975 to the late 1990s, which cost thousands of lives and brought enormous suffering (Levene 1999; Mohsin 2005; Chakma and Hill 2013). It was only after the Peace Accord in 1997 that various national and international NGOs could access the area and promote new forms of agriculture (during the resistance, Pahari peoples had focused on traditional practices). Pahari political élites established their own NGOs to resist national NGOs that privileged Bengali interests. As explained below, these Pahari-led NGOs differed from others in their promotion of traditional practices over profit-oriented activities, although they were not averse to accommodating certain capitalist agendas.

In the late 1990s and early 2000s a number of neoliberal microcredit NGOs such as the Association for Social Advancement (ASA), Building Resources Across Communities (BRAC), the Integrated Development Fund (IDF), Grameen Bank and others, began working for the socioeconomic 'improvement' of CHT communities. A similar agenda had been implemented in the 1980s by government development agencies such as the Chittagong Hill Tracts Development Board (CHTDB) and local livestock and agricultural development extension offices. During this period several international development NGOs and agencies such as the United Nations Development Program (UNDP), the World Food Program (WFP), the Asian Development Bank (ADB) and World Vision (an international NGO) also began to work on the social, economic and infrastructural development of the CHT region. Local agricultural extension offices were another important way in which development organizations introduced new technologies, as these provided subsidized seeds. These institutions continue to implement various development programmes through local partner NGOs and government development agencies.

Neoliberal microcredit NGOs now encourage their programme participants to invest loans in economic production, including agricultural production, in order to produce marketable goods and earn a profit. Some of these NGOs (for example, BRAC) also insist that local farmers adopt new technologies to increase agricultural production. Examples of these technologies include high-yielding variety seeds (HYVs), chemical fertilizers and pesticides, power tillers and rice-husking machines. Indigenous NGOs, such as Green Hill, Taungya and the Centre for Indigenous Peoples Development (CIPD, a partner NGO of UNDP), encourage Paharis to preserve natural resources and biodiversity, but they do not oppose the integration of new technologies with traditional farming practices. One difference between indigenous and neoliberal NGOs is that the neoliberal NGOs emphasize maximizing production on all lands, using technologies to improve agricultural yield, while indigenous NGOs allow fields to remain fallow for at least two years. Indigenous NGOs thus accommodate new technologies while emphasizing traditional values and practices that maintain communal landscapes.

Governmental and neoliberal development agencies and organizations equate the continuation of *jhum* cultivation with high poverty rates among Pahari peoples. An area manager of a local partner NGO of UNDP explained that the CHT rural development program (CHT-RDP) was implemented to improve the social and economic lives of poor Pahari peoples. According to this NGO manager, most Pahari peoples are living in poverty because of their lack of effort to change their way of life. He argued that Pahari peoples must change their traditional worldview, avail to the benefits of modern agricultural technologies and various social and economic development opportunities and increase agricultural production.

> Most Paharis still depend on *jhum* cultivation and what they produce through this traditional farming is insufficient for meeting tomorrow's needs. Based on the availability of food at home they design their next plan of actions [*sic*]. They do not care about other things if they have enough food available at home to feed themselves and their children. But there are many opportunities here to utilize natural [resources] to increase agricultural

production, both to meet their family needs as well as earning money to save for the future. They can raise livestock such as cows, goats, hens and ducks.

(Interview on 28 July 2009)

These sentiments are not uncommon among development NGOs and workers. Connected with the push towards modern technologies is the perceived need to produce cash crops or livestock – that is, to convert traditional subsistence economies into 'modern' capitalist economies.

Initial responses from Pahari farmers were mixed; some consciously rejected such development ideas, some were indifferent and sceptical about the effectiveness of new agricultural technologies, while others began to incorporate those technologies while continuing their traditional *jhum* farming system. One Pahari participant, Bikash Chakma, who had much experience working with development organizations, provided his thoughts on how Pahari reacted to new technologies for livestock hybridization and agricultural productivity. Bikash was a duck procreation project manager under the Rangamati Upazila Livestock Office. Previously he had worked for CIPD. At that time, CIPD was implementing a homestead food production project of the charity organization Helen Keller International and Bikash was responsible for implementing this project among the Pahari farmers in Rangamati. However, few Paharis welcomed the scientific technologies offered to them, as he recalls:

We went door to door to farmers and offered our technologies related to agriculture and homestead food production, but these peoples rarely showed any interest in these technologies. Though some of them received us, they did not use [the technologies] properly. Some of them carelessly put these technologies [e.g. HYVs] in places so that they [the seeds] were wasted or rats ate them. In some places, we faced direct resistance [by] local farmers when we tried to advise them to [shift from] *jhum* cultivation. Some years ago . . . some of the Pahari farmers [even physically] beat our team members when they went to a remote hill in Rangamati to convince them to accept new agricultural technologies such as HYVs, cattle hybridization and other artificial livestock procreation technologies.

(Interview on 30 July 2009)

NGOs and government agencies have used various methods to convince Pahari farmers to accept 'environmentally friendly and productive' agricultural technologies and to cease traditional *jhum* cultivation. In the early 2000s, UNDP started disbursing BDT40,000 (US$515) as grants to each Para [neighborhood] Development Committee (PDC) through local partner NGOs under the UNDP's Capacity-Building and Community Development Project. This programme did affect uptake by some Pahari farmers. For instance, some of them incorporated new technologies such as power tillers, rice husking machines, HYVs, fertilizers and pesticides, as well as new foods including fruits and honeybees.

Figure 2.1 A view of *jhum* land prepared for cultivation after clearing the jungle.

Source: Authors' own photo.

The incorporation of Pahari farmers into capitalist economic food systems is thus advanced by microcredit NGOs, who provide loans for commercial food production but not subsistence farming. However, adoption of new agricultural technologies and practices is not uniform across the CHT region and *jhum* cultivation continues in our research area (see Figure 2.1). Still, there have been complex shifts in how *jhum* is practised. While land was traditionally left fallow to ensure future productivity, many Pahari farmers now use fertilizers and pesticides to increase agricultural productivity in their remaining lands. Other Pahari people claim that land fertility has recently diminished because of this increased use of chemicals, exacerbated by local effects of climate change such as drought. Some prefer *jhum* cultivation as it helps to protect the fertility of land and to preserve natural resources for their livelihoods:

> Previously our peoples [Paharis] used to produce what they needed to survive through *jhum* in the hills. Now many of them are trying to incorporate new agricultural technologies, turning to horticulture or mixed fruits gardening and producing cash crops such as ginger, turmeric, sugarcanes, etc. Most of these activities are responsible for destroying the rare wild plants and the beautiful habitat of wild animals. This is harmful for our peoples, our culture and [our] local economy. Farmers who have large plots of land

might benefit economically from such agricultural practice[s], but there have been many negative effects on [the] local environment, biodiversity and above all, sources of [Pahari] livelihoods. Lack of rain, the reduction of agricultural fertility and loss of rare wild plants and wild animals are some of the key indicators that our biodiversity has [been] affected.

(Interview on 25 June 2009)

As this statement suggests, some Pahari people are concerned with changing production practices as they threaten the preservation of local environments and natural resources.

New agricultural technologies also increase inequalities between Pahari farmers. Indeed, much research has been conducted on the 'double-edged sword' of new agricultural technologies and agricultural commercialization for farming communities across the world (see Shiva 1991; Bernstein *et al.* 1992; Bernstein 2010). Henry Bernstein, for example, demonstrates how the introduction of new agricultural technologies and commercialization in Africa benefits wealthier farmers while forcing landless and marginal farmers to leave farming to search alternative livelihoods, either in rural or in urban areas. Vandana Shiva (1991) argues that the Green Revolution in India left farmers with debt, communal conflict, violence and the destruction of natural resources and ecological systems. The introduction of new agricultural technologies into the CHT has brought similar consequences. Big farmers benefit from increasing agricultural productivity brought by the application of HYVs, chemical fertilizers and pesticides, at the expense of local environments and agroecologies. Small farmers, on the other hand, are encountering various problems in sustaining their traditional livelihoods. Small and marginal farmers, who could produce food previously, cannot afford the increasing costs of inputs such as seeds and chemical fertilizers. Many of these farmers question the idea that modern inputs lead to 'better' food production. As one Pahari farmer said:

Previously our land was fertile and we could grow huge crops naturally to meet our needs. We didn't need to buy seeds every year, to [apply] fertilizers or pesticides, as we have to do today. But now we cannot even think that crops, vegetables or fruits can grow that way. So now we must [apply] fertilizers, pesticides, get the land irrigated and consult with the agricultural extension officers in order to have better food production. Yet there is no guarantee [of] that.

(Interview on 20 June 2011)

In our study, we found that many such farmers had lost their own lands because of household debt and were seeking alternative sources of income by becoming agricultural day labourers, working in textile factories or forming small businesses.

As indicated earlier, some local NGOs are encouraging Pahari peoples to protect the natural environment and forest resources. The local NGO, Taungya, for example, established and owned by the sitting Chakma Chief Circle,[8] is mobilizing Pahari peoples in Rangamati to protect and preserve forests and forest resources such as bamboo, edible plants and trees. Such forest resources are very significant for the livelihoods of Pahari villagers because they are repositories of food, herbs, medicinal plants and wildlife (Roy and Halim 2001). Many Pahari farmers we spoke with said that the preservation of forests and forest resources not only helps to preserve the quality of the soil, but also increases biodiversity. By contrast, NGOs such as CIPD and Green Hill are encouraging Paharis to incorporate new agricultural technologies such as modern plough technology, HYVs and fertilizers in order to increase agricultural production to meet their economic needs – and clearly some Pahari are incorporating these innovations into their farming practices. The influence of these NGOs illustrates the 'alternative-additional' scenario of Fuller and Jonas (2003), in that many Pahari people continue to create spaces and opportunities for traditional food practices and values while selectively choosing to incorporate agricultural technologies.

By contrast, Pahari peoples involved in earlier resistance movements against Bangladeshi armed forces and migrant Bengalis aimed to protect their distinct cultural identity, economic rights (including land rights) and political autonomy against the profit-oriented social norms of the Bengalis. During this time, Pahari peoples opposed hegemonic political and economic ideologies of the Bangladeshi State as well as Bengali cultural values and practices; in this way, their food networks were arguably 'alternative-oppositional'. However, since the late 1990s when the CHT Peace Accord was signed, a shift from an 'alternative-oppositional' to an 'alternative-additional' scenario is evident, as some Pahari peoples selectively adopt Bengali (and international) economic norms and values while continuing traditional food production and distribution practices. Change is evident as some Pahari peoples are becoming involved in agricultural businesses and have access to mobile communication about the price of food in regional and national markets. Some Pahari peoples have now begun to enter the food distribution business, most often in partnership with Bengalis. Below we elaborate our discussion and analysis of such 'alternative-additional' Pahari food networks with ethnographic examples that partly shift our focus from food production to food exchange/distribution and consumption.

Contemporary Pahari food systems

This final section uses ethnographic data to describe how economic and social–cultural factors shape Pahari food production, exchange/distribution and consumption patterns, enabling Pahari peoples to carve out particular 'alternative-additional' spaces for local food and drink such as mushrooms, rice wine and fish.

As noted earlier, traditional Pahari economic activity can be distinguished from that of Bengalis, whose major economic activities centre on agricultural trade. Indeed, the economic activity and cultural values and norms of the Pahari peoples have long been different from that of the majority of Bengalis, and this difference influences the ways food (and drink) is produced, exchanged and consumed. According to our participants, most Pahari farmers in Rangamati were entirely subsistence farmers until the late 1990s, when some began growing cash crops as encouraged by government agricultural extension officers and development agencies. While most Pahari farmers who adopted new foods and technologies continued to produce a variety of local food items alongside new agricultural crops, some focused on less commercially distributed local foods.

Local consumers, both Pahari and Bengali, purchase vegetables, fish, fruits and other food items directly from Pahari farmers in various marketplaces in Rangamati. Bengali traders always look at what local farmers (who are predominantly Pahari) are producing and harvesting. These businessmen (always men) also maintain a very good relationship with the businessmen in Rangamati and Chittagong. It is the Bengali businessmen who control the price of food and cash crops in these districts. Local Pahari farmers feel exploited by them since they argue that they receive lower than market prices from Bengali traders. While it is normal for the market price of agricultural products to vary in local and regional markets, Pahari farmers claim they are deprived of the actual market value because of Bengali domination in the local and regional marketplaces.

We found that locally produced foods are readily available in various marketplaces. These food items and other agro-products are supplied directly to consumers in local, regional and national marketplaces through different means. One example is the cultivation of mushrooms. A mushroom development project ran in Rangamati from 1999 until 2006, resuming again in 2009. Mushrooms were a traditional Pahari food but in this scheme Pahari peoples were also encouraged to incorporate new technologies in order to market their mushrooms. Interested farmers were provided with training and subsidized seeds produced in a laboratory at the Mushroom Development Extension Centre (MDC). Bengali Muslims rarely participated in the project because most Bengalis question whether mushrooms are *halal*, or allowable in Islamic traditions.[9]

According to the project manager of the MDC, while it was difficult for the agricultural officers to incorporate Muslim Bengalis into the mushroom development project, most Pahari farmers showed interest in the production of mushrooms for consumption and for sale at the local market. The project manager explained:

> We have a mushroom development project, which has been designed according to the food habits and marketing demands of mushroom[s] in Rangamati. Whereas in [the] plains districts it takes about six months to make people understand whether mushroom is *halal* or *haram* [forbidden], people here are familiar with it and they are used to eating mushroom as a vegetable. Unlike the Muslim Bengalis, indigenous Paharis have eaten mushrooms for generations and this food practice inspired the large mushroom development project in the CHT region. In the first phase, we are providing training and seeds to local farmers. Our next phase is to create a market for the mushroom products. It is also our objective to create some market opportunities for these farmers and so we are communicating with various stakeholders who are interested in purchasing mushrooms and mushroom products from local markets.
>
> (Interview on 29 July 2009)

In 2011, many Pahari peoples were cultivating mushrooms in their homes, and raw mushrooms and mushroom products were being sold in local marketplaces. Even some of the poor Bengali families had begun cultivating mushrooms. The main buyers of these mushrooms were vegetable vendors and restaurant owners.

The production, sale and consumption of mushrooms may be seen as an 'alternative-additional' food network since the MDC encourages both commercially driven agriculture to improve the economic status of Pahari as well as traditional practices. Such traditional practices include the production of mushrooms at the household level and the continued preparation of *pakura* (mushrooms fried with flour, spices, oil and other items) and of mushrooms cooked with fish or meat. MDC officials encouraged such traditional practices by providing subsidies or free packets of mushroom seeds and free training to local farmers.

Another 'alternative-additional' food system in Rangamati is rice wine production. Producing, sharing and consuming wine is grounded in traditional Pahari food practices. Unlike Muslim Bengalis, Pahari peoples do not have any religious restrictions against drinking alcohol and many Pahari peoples produce rice wine for everyday use and special occasions such as marriage ceremonies, birth rituals and other festivals. While the majority of Pahari peoples are followers of Buddhism, they are influenced by a plurality of religious and spiritual practices that incorporate animism and Hinduism. Among the Chakmas, for example, there is a belief in *Lakshmi*, the goddess of crops, and *Wiya*, the god of house (Khan *et al.* 2004). *Lakshmi* is the Hindu goddess of prosperity, while *Wiya* is a traditional god. Chakmas also celebrate the *Biju*, the Bengali New Year, in order to seek relief from all forms of sins and bad deeds committed over the past year and to have a happy and auspicious new year. During this celebration, guests and visiting relatives are offered traditional food and rice wine.

While Pahari peoples are allowed to produce such wine for consumption within the household or for offering to their guests for celebrating special occasions, the production of rice wine for commercial sale, or purchasing and drinking wine in public places, has been traditionally and legally prohibited. There are only a few government-licensed bars, which are mostly located in the international hotels and diplomatic zones in major cities (for example, Dhaka and Chittagong), where wine can be sold to eligible customers who must follow the government's alcoholic policy.[10] Usually, these wines are imported from other countries and not locally produced.

Despite legal and cultural restrictions, the production, sale and consumption of rice wine has increased in Rangamati because of: (i) the increasing popularity of rice wine in populations of all ethnicities and religions, (ii) the lack of employment and household poverty and (iii) the availability and accessibility of funds from microcredit NGOs. We argue that rice wine production and consumption in the Rangamati is both alternative-traditional and alternative-additional, and that its 'additional' component is a recent phenomenon.

The narratives of our research participants revealed that Pahari peoples had not started to sell rice wine in Rangamati until the late 1990s. Several factors contributed to this, such as lack of employment and poverty associated with the depletion of forest resources, and an increased demand for local wine by both the Pahari peoples and (non-Muslim or unorthodox Muslim) Bengalis, especially of the younger generations. The majority of the customers are Pahari peoples, however, who usually go to the producers' house to purchase the wine because the producer cannot openly sell their wine on the market. Our research showed that, despite its legal prohibition and threat of police raids, many Pahari families are producing rice wine for sale. These Pahari are building upon traditional food (or in this case drink) practices while joining new capitalist markets.

Our final example of 'alternative-additional' foodways examines how (neo) colonialism has created new food market opportunities. Earlier we described the negative effects of the Kaptai Dam on Pahari peoples in Rangamati. However, there is at least one positive outcome of the Dam for the local food economy: the resultant reservoir, called Kaptai Lake, is now a key source of fish for the Pahari. Kaptai Lake has not only shaped local peoples' everyday food consumption practices, but the availability of this food item has also created new spaces for local and regional fish markets. According to the Bangladesh Fishery Research Institute, Kaptai Lake is now the most diversified and prosperous source of water resources, containing a total of seventy-six varieties of fresh water fish, most of which are local species (sixty-eight are local fresh water fish and eight are foreign species). Currently, a significant number of these fish (about forty-two types) are being collected from the Kaptai Lake for commercial purposes (Khan *et al.* 2004). Regardless of economic status, many Pahari peoples and Bengalis depend upon fishing for their livelihoods.

Before the construction of the reservoir in the early 1960s, the Karnphuli River was the only source of fresh water fish for Pahari peoples. Like Bengalis, Pahari peoples traditionally consumed a variety of fish, especially those people living in the Karnphuli River Valley area (see Figure 2.2). However, there are some fish and fish dishes (for example, *nappi*, see Figure 2.3) that only Pahari people eat. Bengalis avoid these fish and fish dishes because of their doubt about whether they are *halal*. *Nappi* is the most popular traditional fish dish for Pahari peoples and it is also an essential food ingredient. It is made of various rotten and dried fish and seafood (such as shrimp and very small types of local fish) and snails. To prepare *nappi*, various small fish, shrimp and snails are put into a pot to dry out for a few days, after which they are ground into a paste. Most Pahari people use *nappi* with which to cook vegetables, fish or meat curries and to prepare many other everyday food items. *Nappi* works like cooking oil, bringing a special flavour and taste to food. Many Pahari peoples believe that *nappi* improves the immune system, protecting people from illnesses caused by mosquitos, snakes and poisonous insects that live in the forest. This food item serves to mark cultural boundaries between Pahari peoples and Bengalis and thus symbolizes a unique Pahari food identity.

Traditionally, Pahari peoples in Rangamati fished for subsistence purposes. According to the narratives of our research participants, Chakmas living in the river valley area and those in the hills caught fish from the river, small water streams or the canal for everyday consumption and for sharing with relatives and

Figure 2.2 A fisherman is busy in the Kaptai Lake/Karnphuli River Valley area in Rangamati.

Source: Authors' own photo.

neighbours, rather than for sale. The construction of Kaptai Lake greatly reshaped fish consumption patterns and sources of livelihood as hundreds of canals and wetlands surrounding the lake provided new sources for fish. Since the post-CHT Peace Accord period in the late 1990s, some Pahari and Bengali families have become engaged in community fishery development projects, which involves limited investment, with farmers providing feed and then harvesting fish when they have grown to an adequate size. Fishing now provides for household subsistence as well as income. Pahari peoples' intensive engagement in fishing, either in individual or community-based fishery projects, is 'alternative-additional' because many Pahari peoples are using this opportunity to make a profit, but also to foster traditional subsistence foodways such as *nappi* production and consumption.

The president of a Pahari community fishery project explained how the fishing business worked:

We have recently started a fishery project and I am the president of this project [see Figure 2.4]. There is another fishery project beside ours, which is owned by our *Karbari* [traditional Pahari village/community leader]. I could do my own fishery project too, but I don't want that some people will eat and

Figure 2.3 A trader selling *nappi* in Rangamati town.

Source: Authors' own photo.

Figure 2.4 A view of a community fishery project in the research village in Rangamati.

Source: Authors' own photo.

other[s] would die [*sic*]. So I make space for people from every household in our neighborhood and allow those who are interested in participating in my project. We name this project 'Mittingachari Youth and Fish Development Association.' . . . We have started this project by borrowing Tk. 70,000 from a Bengali fish merchant. We needed this money to purchase fishnets, feed and other things. So far we have caught [a harvest of] fish two times from this canal and have paid Tk. 20,000 to the fish merchant. It is the condition that whatever amounts of fish we can catch we have to sell to this merchant even after we repay all our debts to him. He will give us the price of fish according to wholesale market prices. When [we] repay this debt we will be able to establish our common ownership over fishnets and all incomes from this project.

(Interview on 2 March 2011)

This brief quotation suggests that Pahari peoples are now engaging in community-based fishery projects for subsistence as well as profit. Indeed, according to this research participant, the Mittingachari Youth and Fish Development Association project provides a means of subsistence for himself as well as for low-income people within his community. Though some members of the community profit from such projects, all Pahari peoples share resources for their livelihoods, even in changing socioeconomic and political contexts.

Like fish, locally grown foods are used for household and community consumption prior to sale. Surplus food items such as ginger, turmeric, pineapples, litchis, jackfruits and bananas are sold in local marketplaces. Among Pahari people who participate in the marketplace, there is still an emphasis on subsistence in addition to profit-maximization. Non-economic factors such as Pahari peoples' 'passion, tradition and identity' play an important role in shaping Pahari food production, consumption and economic activity (see also Trefry *et al.* 2014). A Pahari elderly man said:

You know that we, the people here in the CHT, are different from that of the plains. For instance we, the Pahari peoples, don't feel comfortable if we don't go to our hut in the *jhum* land at least once a day because it is not just our occupation but is also our passion, our tradition and identity. We are habituated to go there from our forefathers, grandfathers and father. I have planted ginger or taro roots or I make a garden of mixed fruits such as banana, watermelon and so I have to go there at least once a day. But the majority of plain districts people work in a farming land do business or do job[s] in offices, which is significantly different from us.

(Interview on 2 March 2011)

Our ethnographic data demonstrate that postcolonial foodways in the CHT produce 'local distinctiveness' (Fuller and Jonas 2003: 70), as Pahari peoples continue to privilege traditional economic values and food production practices, while also integrating in local and regional food markets. Despite increasing focus by some local farmers on commercial food and agricultural production and integration by Pahari peoples into wider market systems, it is clear that many Pahari peoples are maintaining their traditional food practices by continuing to participate in sharing economies and prioritizing familial and community subsistence and ritual needs. In other words, while integrating somewhat into marketplaces, Pahari peoples continue to contest dominant food and agricultural norms by privileging subsistence in addition to profit, reproducing 'alternative-additional' economic spaces. Foods that Pahari peoples eat, with the exception of cooking oil, salt and other spices, are produced on their lands, often by traditional *jhum* agriculture. From this perspective, they have little dependency on external food markets.

Conclusion

Local food systems in CHT are contested and shaped by multiple factors: (i) colonial and postcolonial economic and development policies; (ii) pressures from government development agencies and some NGOs to cease *jhum* cultivation and to adopt new agricultural technologies; (iii) local NGOs' emphasis on the protection of biodiversity and local environmental sustainability and, most importantly, (iv) the continued importance of food sharing and exchange, as well as many Pahari peoples' reluctance or inability to fully engage with profit-oriented food production.

Using ethnographic data, in this chapter we have shown how Pahari farmers who had been living on subsistence agriculture for centuries responded to pressures from colonial and postcolonial governments and development NGOs to adopt new agricultural technologies and maximize profits. Although local Pahari farmers adopt some aspects of modern food production and marketing systems, local foodways differ from national and global food systems in key ways. Our findings show that traditional subsistence practices still dominate Pahari food economies with selective incorporation of hegemonic foodways. The continuation of *jhum* cultivation by Pahari farmers, despite pressures to leave this practice, suggests that many Pahari peoples consciously resist hegemonic modes of production and want to develop a different food system. Moreover, the integration of Pahari farmers and businessmen into local and regional food markets does not necessarily suggest that local foodways and traditional values are shifting towards capitalistic values. Instead, such integration is creating possibilities for new economic and cultural relationships between Pahari farmers, traders and consumers, only some of which are tied to profit-making.

Since Pahari peoples are incorporating some elements and norms of capitalism, adopting some new agricultural technologies and focusing on market expansion for local food, in addition to continuing traditional practices of food production, food sharing and food preparation and consumption, we have argued that this postcolonial food system can best be understood as an 'alternative-additional' food network. Pahari peoples are not a singular entity with one voice, and some have tried to acquire business skills from Bengalis, despite their historical and recent antagonistic relations. Nevertheless, the majority of Pahari peoples continue to use food as a means of ascertaining their identity, replicating traditional food production and distribution practices and maintaining Pahari social values. Future research in the area could reveal how alternative-additional foodways continue to evolve in Pahari communities, as relationships between mainstream and alternative/traditional economic spaces continue to change in the Chittagong Hill Districts of modern Bangladesh.

Notes

1   These ethnic groups are the Chakma, Marma, Tripura, Tanchangya, Khumi, Khyang, Mru, Lushai, Chak, Pankua, Bawm, Ahamiya and Chhetri.
2   These ethnic groups are the Chakma, Marma, Tripura, Tanchangya, Lushai, Ahamiya and Chhetri.
3   The colonial period ended in 1947 when the British left the Indian sub-continent and India and Pakistan became two sovereign countries.
4   After violent altercations with Paharis, The East India Company subjugated their leaders, creating a tributary state in the CHT (Gupta 2005).
5   All names of research participants are pseudonyms, but their ethnic identity is indicated.
6   Baishakh is the first month of the Bengali New Year and the first day of the month is 14 April.
7   Jamuna bridge was constructed over the Jamuna River between the years 1994 and 1998. The second longest bridge in South Asia (at 4.8 kilometres), it is located 110 kilometres northwest of Dhaka, the capital city of Bangladesh. The World Bank, Asian Development Bank, the Government of Japan and the Government of Bangladesh financed the construction of this bridge. During the late 1980s to the early 1990s, thousands of Bengali households, who were mostly farmers, had been displaced from both sides of the Jamuna River. Various studies show that many of these families had to migrate to other places involuntarily, including Dhaka, Chittagong and the Chittagong Hill Tracts (see also Rahman 2010).
8   Chakma Chief Circle refers to the area that the Chakma hereditary leader controls. The leader (chief) has a counsel consisting of local leaders that provide advice on political, social and economic issues in the area.
9   *Halal* can refer to any food that is religiously acceptable to eat, not just in reference to meat. Debates about whether a food is *halal* or not also involve the consumption of shellfish, which some Muslims see as non-*halal* (or *haram*).
10   While there are official restrictions for sales of alcoholic beverages, people with foreign passports may consume alcohol at designated hotel bars or can carry a very limited amount of alcohol on their person. For more details see: www.who.int/ substance_abuse/publications/global_alcohol_report/profiles/bgd.pdf (accessed on 2 September 2015).

References

Ali, A. (2012) 'Place and contested identity: Portraying the role of the place in shaping common sociopolitical identity in Chittagong Hill Tracts, Bangladesh'. *Diversipede* 1(2): 31–46.

Ashcroft, B., G. Griffiths and H. Tiffin (1998) *Key Concepts in Postcolonial Studies.* London: Routledge.

Bhabha, H. (1994) *The Location of Culture.* London: Routledge.

Bernstein, H. (2010) *Class Dynamics of Agrarian Change.* Halifax, Canada: Fernwood Publishing.

Bernstein, H., B. Crow and H. Johnson (1992) *Rural Livelihoods: Crises and Responses.* Oxford: Oxford University Press.

Borggaard, O. K., A. Gafur and L. Petersen (2003) 'Sustainability appraisal of shifting cultivation in the Chittagong Hill Tracts of Bangladesh'. *Ambio: A Journal of the Human Environment* 32(3): 118–23.

Chakma, K. and G. Hill (2013) 'Indigenous women and culture in the colonized Chittagong Hill Tracts of Bangladesh'. In Kamala Visweswaran (ed.) *Everyday Occupations: Experiencing Militarism in South Asia and the Middle East.* London: SAGE Publications, 132–57.

Childs, P. and P. Williams (1997) *An Introduction to Post-Colonial Theory.* London: Prentice Hall.

Fuller, D. and A. Jonas (2003) 'Alternative financial spaces'. In Andrew Leyshon, Roger Lee and Colin Williams (eds) *Alternative Economic Spaces.* London: SAGE Publications, 55–73.

Gopal, T. and G. Rasul (2005) 'Patterns and determinants of agricultural systems in the Chittagong Hill Tracts of Bangladesh'. *Agricultural Systems* 84(1): 255–77.

Gupta, A. (2005) *Human Rights of Indigenous Peoples. Volume 2.* Delhi: Isha Books.

Harvey, D. (2005) *A Brief History of Neoliberalism.* Oxford: Oxford University Press.

——(2007) 'Neoliberalism as creative destruction'. *The ANNALS of the American Academy of Political and Social Science* 610(1): 21–44.

Jonas, A. (2010) '"Alternative" this, "alternative" that . . .: Interrogating alterity and diversity'. In Duncan Fuller, Andrew Jonas and Roger Lee (eds) *Interrogating Alterity: Alternative Economic and Political Spaces.* Burlington, VT: Ashgate, 1–27.

Khan, Z., S. Sikder, J. Alam and A. Dey (2004) *Rangamati: Boichitrer Oikatan (Harmony of Diversity).* Rangamati: District Administration of Rangamati.

Levene, M. (1999) 'The Chittagong Hill Tracts: A case study in the political economy of "creeping" genocide'. *Third World Quarterly* 20(2): 339–69.

Leyshon, A. and R. Lee (2003) 'Conclusions: Re-making geographies and the construction of. "spaces of hope"'. In Andrew Leyshon, Roger Lee and Colin Williams (eds) *Alternative Economic Spaces.* London: SAGE Publications, 193–8.

Loomba, A. (1998) *Colonialism/Postcolonialism.* London: Routledge.

Majumder, S., K. Bala and A. Hossain (2012) 'Food security of the Hill Tracts of Chittagong in Bangladesh'. *Journal of Natural Resources Policy Research* 4(1): 43–60.

Mallick, D. and M. Rafi (2010) 'Are female-headed households more food insecure? Evidence from Bangladesh'. *World Development* 38(4): 593–605.

Mohsin, A. (1997) *The Politics of Nationalism: The Case of Chittagong Hill Tracts, Bangladesh.* Dhaka: The University Press Limited.

——(2005) 'Globalization, hegemony and the State in Bangladesh'. In Patricio Abinales, Ishikawa Noboru and Tanabe Akio (eds) *Dislocating Nation-States: Globalization in Asia and Africa.* Kyoto: Kyoto University Press, 64–88.

Nandy, A. (1983) *The Intimate Enemy: Loss and Recovery of Self under Colonialism.* New Delhi: Oxford University Press.

Nasreen Z. and M. Togawa (2002) 'Politics of development: "Pahari–Bengali" discourse in the Chittagong Hill Tracts'. *Journal of International Development and Cooperation* 9(1): 97–112.

Rahman, M. (2010) 'Impact of riverbank erosion hazard in the Jamuna floodplain areas in Bangladesh'. *Journal of Science Foundation* 8(1 and 2): 55–65.

Rasul, G. (2007) 'Political ecology of the degradation of forest commons in the Chittagong Hill Tracts of Bangladesh'. *Environmental Conservation* 34(2): 153–63.

Roy, R. and S. Halim (2001) 'Valuing village commons in forestry: A case from the Chittagong Hill Tracts'. In Quamrul Chowdhury (ed.) *Chittagong Hill Tracts: State of the Environment*. Dhaka: Forum of Environmental Journalists of Bangladesh, 11–44.

Schendel, W. (1992) 'The invention of the "Jummas": State formation and ethnicity in Southeastern Bangladesh'. *Modern Asian Studies* 26(1): 95–128.

Shiva, V. (1991) *The Violence of the Green Revolution: Third World Agriculture, Ecology and Politics*. London: Zed Books.

Trefry, A., J. Parkins and G. Cundill (2014) 'Culture and food security: A case study of homestead food production in South Africa'. *Food Security* 6: 555–65.

Tripura, S. (2008) 'Blaming Jhum, denying Jhumia. Challenges of the Jhumia indigenous peoples land rights in the Chittagong Hill Tracts (CHT) of Bangladesh: A case study on Chakma and Tripura'. M.Phil. thesis in Indigenous Studies, the Faculty of Social Sciences, University of Tromsø, Norway. http://munin.uit.no/handle/10037/1535 (accessed 4 January 2015).

UNDP (United Nations Development Program) (2009) *Socio-Economic Baseline Survey of Chittagong Hill Tracts*. Dhaka: Human Development Research Centre.

3 Justice for the salmon

Indigenous ways of life as a critical resource in envisioning alternative futures

Sophia Woodman and Charles R. Menzies

We are the salmon people
Farmed salmon kills our way of life
Nuxalk Heiltsuk Ulkatcho

Protest banner, 2003, Vancouver[1]

Since the 'Idle No More' protest movement sparked a new wave of indigenous activism across Canada starting in December 2012, 'justice for the wild salmon' has been a central theme of some of the associated actions, including marches, petitions and lobbying efforts. This theme has also emerged in a transnational campaign against the Norwegian multinationals that are the dominant actors in the global industrial farming of salmon. Some of these campaigning activities emerge from collaborations between indigenous activists and environmentalists and scientists concerned about the impact of salmon farms, in the Canadian Pacific coast and elsewhere, on depleted stocks of the wild fish. Activists and some scientists assert that the rapid growth of Atlantic salmon aquaculture in this region, along with the long-term deleterious effects of resource extraction and hydro projects on salmon habitats, have contributed to a precipitous decline in numbers of salmon returning to spawn in the rivers that flow into the Pacific. These effects have been so severe that the Canadian government set up a Commission of Inquiry in 2009 to investigate the declining run of salmon in the Fraser River.

Soon after the launch of the 'Idle No More' movement, a 'March for Wild Salmon' was organized by the Indigenous Salmon Defenders in Vancouver, British Columbia. This march was one of a global chain of events focused on the threat to wild fish posed by salmon aquaculture, with the message: 'Stop Norwegian fish farms from killing wild salmon!' The first demonstration held on 1 March 2013 was a march to the Vancouver offices of the Canadian Department of Fisheries and Oceans, which licenses fish farms. Organizers situated their action in an alternative space by noting that they were marching on the territory of the Coast Salish people (see Figure 3.1). On the same day, a parallel event was staged in Oslo, Norway and another in Galway Bay in Ireland, where a major expansion of fish farms was being planned. A few days later, on 6 March, a

Figure 3.1 Salmon farms of coastal British Columbia.

Source: Map produced by Living Oceans (www.livingoceans.org).

demonstration was staged in Bergen, Norway, and organizers also planned events in Edinburgh that month.[2] The march in Canada was one of many such actions that have continued over the last few years, most recently with the 'Caravan for Wild Salmon' in British Columbia in May 2015, described in more detail below.

It is no accident that these actions began in British Columbia, and that indigenous activists were central to framing what is at stake. While this is a transnational campaign against industrial salmon farming that connects groups in various countries against a perceived common threat, the social manifestations of this contention have been distinctly different in each locality. This chapter explores the reasons for these differences, focusing on three interrelated areas: the impact and socio-economic basis of indigenous systems of knowledge that are sustained by land- and water-based practices; the connection of such practices to collective claims to land, rivers and sea that challenge the persistence of the colonial order; and the crucial role of the state in creating the conditions for contention over salmon farms, by imposing frameworks of law and property relations and promoting particular forms of capital accumulation.

Through comparing struggles around salmon farming in Europe and North America and in the transnational campaign, this chapter highlights the distinctive work done by indigenous activists in making justice for the salmon central to the struggle. Indigenous visions evoke 'new modes of place-making' (Soguk 2007: 22) that are simultaneously rooted in localities and transnational in scope, generating resources for a radical politics that asserts the possibility of a different world. Going beyond the competing versions of science evident in many environmental struggles, indigenous campaigns against industrial salmon farming assert a fundamental challenge to the hegemony of modern scientific knowledge in ways comparable to those discussed in Chapters 4 and 7 in this volume. They also assert alternatives to the regime of private property rights that is increasingly dominant around the world. Thus indigenous knowledge and practices can be crucial resources in reclaiming the global commons and protecting and/or rediscovering truly sustainable ways of living in our finite world.

At the end of the chapter, we argue that our comparison calls into question the 'politics' of Actor Network Theory (ANT) accounts of contestation around salmon farms (Lien and Law 2011; Law and Singleton 2013; Lien 2015). ANT accounts seek to problematize the 'modern' divide between nature and culture, but pay insufficient attention to the history of how particular networks involving salmon came to be formed. We show how ANT's foregrounding of technology leaves many connections involved in the salmon farms unexplored, and we argue that such a concentration involves political choices about what to study that are rarely made explicit, while the choices ANT scholars have made in studying salmon can result in accounts that elide questions of justice or sustainability.

This chapter is based on the combined perspectives of the co-authors, and their varied data sources. As a member of the Gitxaała First Nation, Charles Menzies has been engaged in fisheries on the Canadian Pacific coast since childhood. In conjunction with this experiential knowledge, as an anthropologist he has focused his ethnographic work on various dimensions of fisheries in British Columbia, while also working in Brittany and other parts of France. After moving from Vancouver to Edinburgh, Sophia Woodman was struck by the contrast between the contention over fish farms in British Columbia and the lack of similar actions in Scotland, and began to explore the reasons for this by combining data from online materials and related social movements and secondary sources.

Conflicts over salmon

Indigenous people have struggled to preserve the knowledge and practices that provide a basis for sustainable ways of living with the salmon, despite centuries of colonial efforts to disrupt and destroy them. Their connections with the land and sea – as well as with each other – have been central to the survival of these practices (Colombi and Brooks 2012). By contrast, in places such as Scotland and

Ireland, centuries of disruption of connections to land and sea in the name of private property rights have all but expunged analogous knowledge and practices. In the European locations, contention over the impact of salmon farms on wild salmon populations has tended to divide communities, pitting environmentalists against those concerned about jobs and economic activity in areas that are often short of both. Such contests often play out through competing élite visions, generally based on claims about scientific evidence. Similar contests over science are also present in the Canadian context – and some First Nations[3] have accepted salmon farms on their territory as a way of generating local employment and economic development (Gerwing and McDaniels 2006; Brattland and Schreiber 2011). However, the ways indigenous activists of the Canadian Pacific coast include the salmon in the visual and textual representations of their movement highlight a distinctive vision, one of 'justice' for the salmon that enmeshes the fish in a discursive and moral formation grounded in interdependent (yet changing) ways of life in which people have lived with the salmon for thousands of years. Here, living practices, culture, ecology and right to livelihood are intimately bound together in a seamless web. The preservation of a cultural system and an ecological environment are seen as one and the same.

Such a perspective is reflected in the broader vision of the 'Idle No More' movement, which 'revolves around Indigenous Ways of Knowing rooted in Indigenous Sovereignty to protect water, air, land and all creation for future generations.'[4] Colonial projects did not recognize indigenous people as possessing any valid knowledge, and sought to enforce a comprehensive cultural transformation by 'excluding indigenous beliefs, practices and values from the panoply of what could be considered truthful and useful knowledge' (Samson 2013: 69). Today, however, efforts to revive traditional practices and knowledge associated with them are taking indigenous peoples back to the land and water in a variety of different settings across North America (Samson 2013). Through such efforts, indigenous people enact alternative ways of life that resist the continuing effects of colonization and attempts at cultural genocide. The remarkable persistence of such resistance despite centuries of efforts to eliminate indigenous people from colonized spaces 'paradoxically affords them a greater capacity to think imaginatively about possible alternative futures' writes Soguk (2007: 5).

Disputes over fishing methods and fishing rights between indigenous people of the Canadian Pacific coast and first settlers and later, Canadian government agencies, have been endemic since the arrival of Europeans in the region (Newell 1993; Menzies and Butler 2007). Where aboriginal fishing rights were recognized, these were assumed to be only the right to fish for food, rather than to engage in commercial fisheries (Newell 1993). The salmon canneries that became a core industry on the coast were often built on the basis of existing indigenous fisheries, initially depending on their supplies of fish and the labour of indigenous people for both fishing and processing. However, as fishing became subject to government licensing, the customary position of indigenous fishers – who had been involved in sustainable salmon harvesting for their own use and for trade throughout the region for centuries (Campbell and Butler 2010) – was progressively undermined.

Settler-owned canneries and commercial fishing ventures were granted licenses for types of fishing and ownership rights over sites that had traditionally been controlled by First Nations groups. Despite this trend, canneries and fishing camps were also sites of integration between customary practices and the emerging capitalist economy (Menzies and Butler 2007).

This integration did not last. Ironically, perhaps, de-industrialization and post-Fordist modes of production undermined indigenous livelihoods based on catching and processing fish, and indigenous fishers were increasingly pushed out of the commercial fishery. For example, the shift from canned to fresh fish in the mid-1970s meant that fresh-frozen fish moved from shore to vessels, eliminating on-shore jobs. A combination of bans on traditional modes of fishing and substitution of First Nations workers with settlers, as well as a growing emphasis on extraction of resources beyond fish, left coastal First Nations with few opportunities. Limitations on licenses for salmon fishing, introduced as part of a fisheries restructuring plan in 1968, along with the centralization of fish processing, led to precipitous declines in employment in First Nations communities, both as independent fisherpeople and hired workers. As well as affecting labour, regulations reduced fishing units and area licensing, contributing to the concentration of fishing related businesses in ever fewer hands in the 1990s (Menzies and Butler 2007). While all these trends reduced the need for labour, they had particularly severe effects on the First Nations communities that had been almost exclusively reliant on fisheries. In the Gitxaała community, for instance, the combination of traditional marine harvesting and jobs in new economic sectors provided full employment up until the 1960s, but by the 1990s unemployment had become widespread (Menzies and Butler 2007). A similar history is evident in fisheries beyond salmon, such as the harvest of abalone (Menzies 2010). This historical experience has parallels in other chapters in this volume, such as Chapter 1, highlighting how indigenous communities are not necessarily seeking to return to a pre-capitalist subsistence economic system but wish to assert sovereignty over their own food and other resources, and challenge their past exclusion.

Indigenous activists in Canada do not necessarily reject salmon farming per se, but focus on campaigning against a certain variant, one that damages conditions for the uncaged fish, and does not make use of traditional knowledge built up over centuries of living with the salmon. A study of opinions among First Nations communities, some of which have already made deals to have salmon farms on their territory, found that the majority were sceptical about the extent to which expanding aquaculture would address the long-standing problems in their communities (Gerwing and McDaniels 2006). Among community elders and hereditary chiefs of these communities, opinions were 'fundamentally opposed to salmon aquaculture', and tension between these views and those of elected leaders who supported the industry were on the rise (Gerwing and McDaniels 2006: 269).

Resistance to indigenous exclusion from the fisheries has taken various forms, including 'fish-ins' and other types of civil disobedience that asserted indigenous treaty rights over resources such as salmon. In the United States, indigenous fish-ins became a strategy of the Red Power/American Indian Movement from the 1960s onwards, sometimes becoming high-profile events when well-known actors, such as Marlon Brando, joined in (Coates 2006; Wilkes 2006). In Canada, fish-ins have occurred both on the Atlantic and Pacific coasts, involving indigenous members as a group consciously violating federal or provincial fishing regulations as a challenge to the legitimacy of such rules (Wilkes 2006).

In the British Isles, enclosures, disruption of livelihoods and exclusion of small fishers analogous to those in Canada occurred much earlier, and the establishment of rights to private property and limits on the commons were central to creating the framework to legitimize such processes. In Scotland and Ireland over the last 500 years, indigenous residents were repeatedly displaced, disrupting their relations with the land and sea, and destroying collective forms of farming and fishing. Scottish Highlanders were expelled from their traditional territories by their own kin, as clan leaders became an absentee landed class (Hunter 2000). Clansmen and women became small tenant 'crofters' on marginal coastal land after being expelled from their land in the interior by landowners who took it over for sheep farming (Phyne 1997).

In Scotland, salmon has been the exclusive property of the crown since the twelfth century, making the capture and conservation of the fish the subject of some of the earliest known regulations governing natural resources (Coates 2006).[5] Since the mid-1700s, the Crown Estate has owned all of the Scottish coastline and associated marine environments (Phyne 1997). Recognition of customary rights to common land, rivers or marine environments in Scotland only emerged long after the peasant population had been pauperized by the eighteenth- and nineteenth-century shift to commercial landlordism (Hunter 2000). After displacement, coastal herring fisheries became an important source of livelihood for Scottish crofters. However, these fisheries were controlled by landlords, and crofters were forced into employment in herring fishing and in the kelp industry[6] due to inadequate land holdings and high rents (Hunter 2000). Crofting holdings were always too small to maintain a family, and inadequate grazing land limited traditional reliance on pastoralism, so seasonal employment was usually required to sustain livelihoods in the region. This was in part the intention of landlords, who wanted to ensure a ready army of workers for their various ventures (Hunter 2000).

From the 1970s, salmon aquaculture was developed by the Highlands and Islands Development Board as a strategy for bringing economic activity to what was considered a depressed region (Bryden and Scott 1990; Phyne 1997). However, the combination of repeated displacement and centralized ownership by the Crown Estate left coastal people with few opportunities to bring legal challenges against uses of marine resources. Although a similar structure of ownership pertained in Ireland, the advent of the Republic and the subsequent nationalization of some marine resources provided local people with more opportunities to contest measures that diminished their traditional access to coastal waters, with the result that aquaculture firms have often been forced into negotiation with local communities (Phyne 1997).

This is not to say that the native people of Scotland and Ireland always accepted these property relations without resistance. 'Land wars' broke out in Ireland and the Highlands and Islands in the late nineteenth century, focused on winning security of tenure (Hunter 2000). However, the grounds for claiming collective rights had already been so eroded that these struggles concentrated on individual rights to security of tenure.[7] But a vestigial sense of the right to the commons can be observed in popular challenges to the restrictions on access to the resources of land and water, as reflected in the Gaelic saying, 'A deer from the hill, a stick from the wood, or a salmon from the river is no shame to any man'[8] (see also Hunter 2000: 156). The supposed rights of landlords to exclusive access to salmon fishing had to be policed by armies of game-keepers and local sheriffs and constables. Poaching of salmon from rivers in Scotland, England and Wales often involved gangs of men and was aimed in part at challenging aristocratic prerogatives. Nocturnal poaching expeditions were memorialized in poems and songs that celebrated the illegality of the enterprise (Coates 2006).

The dispossession and displacement of small-scale fisher peoples by industrialized fleets and multinational corporations in Canada, Scotland and Ireland reflect a pattern seen across the globe. According to a report by non-governmental organizations (NGOs) including the World Forum of Fisher Peoples, such dispossession has led to a global 'ocean grab' that resembles the earlier privatization of land, with both reducing the global commons (World Forum of Fisher Peoples 2015). While these trends are an aspect of long-term processes – particularly evident in the Scottish case – in recent years, fishing and oceans have become new frontiers for exploiting the global commons for the purposes of capitalist accumulation. Policies that benefit major multinationals and corporate interests are being pursued by states and global actors such as the World Bank in the name of 'conservation', with the rationale that private ownership of marine resources is more efficient and sustainable than local fisheries. As in Scotland and Canada, these policies are promoted in the name of 'development' in the global South. Industrial-scale aquaculture is a key part of such 'solutions' and is said to respond to a global demand for fish to feed an expanding world population (World Forum of Fisher Peoples 2015).

The rise of industrial aquaculture

Although aquaculture has existed in various forms for thousands of years, its use as an industrial farming method is fairly recent. According to data collected by the UN Food and Agriculture Organization (FAO), the sector has expanded exponentially since the mid-1980s, and by the late 2000s, around 30 per cent of global sales were of farmed fish (Gross 2008; Lien 2015). The growth in commercial salmon farming has contributed substantially to this global rise, with annual farmed production increasing from 0.6 million tonnes to two million tonnes between 1990 and 2010. By 2008, about half of salmon sold to consumers was farm-raised (Gross 2008). As with other types of fisheries, small-scale producers have been increasingly pushed out of the business of salmon aquaculture, with one-fifth of the world's farmed salmon produced by just one company,

Norwegian Marine Harvest (World Forum of Fisher Peoples 2015). Such concentration is evident in British Columbia, where three Norwegian multinationals control 92 per cent of the salmon farms (Gillis 2015).

Commercial farming of salmon began in Norway in the 1960s, and gradually spread to other regions amenable to raising fish in captivity (Coates 2006; Page 2007). The main locations for salmon farms are Norway, Scotland, Ireland, Canada, Chile, the Faroe Islands, Japan, New Zealand and Tasmania (see Figure 3.2).[9] The principal method of farming involves hatching salmon eggs and growing the fish in tanks for between twelve and eighteen months, and then releasing the young smolts into net cages where they are raised to maturity in the open sea (Seafood Choices Alliance 2005).

The impact of this form of intensive farming on non-domesticated salmon populations has been very controversial, particularly on the Canadian Pacific coast, where industrial salmon farming began to be practised in the 1970s. Salmon stocks in the Columbia River Basin had already been decimated by resource extraction and hydrological diversions of rivers before the fish began to be farmed there. By the 1990s, the annual run of salmon in the area had fallen to about one million, in comparison to ten to fifteen million returning to spawn there in the early twentieth century (Ladd 2011). While aquaculture advocates claim intensive farming has taken pressure off declining populations of wild fish, farming has

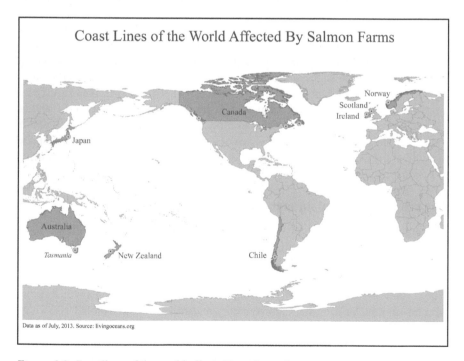

Figure 3.2 Coastlines of the world affected by salmon farms.

Source: Map produced by Living Oceans (www.livingoceans.org).

further threatened their habitat through such impacts as pollution and disease. One threat is the spread of fatal pests and diseases that are endemic in farmed fish, most notably sea lice, a flesh-eating parasite that eventually kills its host. As net cages are entirely permeable, the chemicals used to control such parasites and diseases leach out into the surrounding waters. Another factor is overfishing of the smaller fish on which the carnivorous salmon feeds. For every pound of salmon produced through farming, feed pellets reduced from several pounds of smaller fish, generally wild-caught for this purpose, are required (Lien 2015). Generating sufficient feed for farmed salmon thus reduces the food supply for wild fish. Additionally, waste from the salmon farms is released directly into the sea without treatment, with one farm of 200,000 salmon producing as much faecal matter as an urban centre with a population of 62,000 people (Seafood Choices Alliance 2005). Finally, farmed fish regularly escape from the sea cages, and this is of particular concern where a non-native species is being raised, as is largely the case in aquaculture in British Columbia where the Atlantic salmon is the dominant type of farmed fish (Drews 2013).[10] In 2015, the US Food and Drug Administration (FDA) approved genetically engineered salmon for sale to consumers, clearing the way for extensive farming of this variety of salmon.[11] Most of the salmon farms are located in places where migratory salmon pass on their way to their spawning rivers and creeks (see Figure 3.1) (Schreiber and Newell 2006).

The science around the impact of fish farms on the Canadian Pacific coast is the subject of ceaseless debate among environmentalists, scientists, governments and fish-farm advocates.[12] In the late 2000s, numbers of salmon returning to spawn in the Fraser River – which encompasses a huge watershed incorporating a large proportion of the province of British Columbia – fell so low that a Commission of Inquiry into the Decline of Sockeye Salmon[13] in the Fraser River was established in 2009 by the Canadian government, headed by Supreme Court Judge Bruce Cohen. Over several years, this Commission undertook a comprehensive review of the evidence and policies involved in fisheries and conservation, including the impact of salmon farms on wild fish. While the Commission's findings highlighted a number of threats to salmon – some related to climate change and warming oceans – it identified aquaculture as a significant threat in itself, one that should be addressed as a 'precautionary principle' (Griggs 2012).

Science, property and the state: Scotland and Ireland

Environmentalists have been at the centre of campaigns against salmon farming in Scotland and Ireland. However, much less has been written about these movements than about anti-fish farming campaigns on the Canadian Pacific coast – and there have been no government processes analogous to the Cohen Commission to bring these issues into the public eye. In European settings, disputes over salmon farms are often conflicts between those who support economic development (especially jobs in salmon farms) and environmentalists, sport fishers, tourism operators and local communities concerned about the environmental impacts of industrial aquaculture.

These contests often involve competing versions of scientific evidence, such as: the impact that fish farms have on the numbers of wild salmon that return to particular rivers to spawn; piscine diseases and parasites in farmed fish and their spread to wild fish; and the safety of chemicals used in fish farming on human health. In contrast to the way that salmon is deployed as a participant in the struggles of indigenous peoples in Canada, discussed below, stories and rituals are not apparent in these campaigns, although in the Celtic mythology common to Ireland and Scotland, stories about the 'salmon of knowledge' are a prominent theme, particularly in schoolchildren's education.[14] The connection between these stories that evince reverence for the salmon and traditional practices of conservation have virtually disappeared from living memory in the British Isles.

In Ireland, controversies over salmon farms have emerged periodically since 1987, and have resulted in 'biological warfare', which pits advocates of local development against promoters of tourism, and environmentalists (Phyne 1996: 1). For example, following the 1989 fall of sea trout stocks in rivers in Western Ireland, owners of rivers and lakes used by angling tourists took legal action against salmon farms, leading to years of controversy over scientific evidence of the reasons for the precipitous decline in sea trout. According to Phyne (1996), this was largely a contest between élites, as different state agencies aligned with each side depending on their institutional mandates and relations with the parties involved. Overall, the varying interpretations of science played out against 'a background of different social interests disputing whether the commodification of coastal and inland Ireland should be prioritised in favour of industrial or leisure activities' (Phyne 1996: 20). Here the small-scale fishers were virtually outside the frame, although the outcomes would deeply affect their livelihoods (Phyne 1996). Similar dynamics can be observed in recent disputes over a vast expansion of salmon farms in Galway Bay proposed in 2014 (Cahill 2014).

Anti-salmon farming campaigns in Scotland have some analogous features. An account of a campaign against the siting of new aquaculture farms in Scoraig, a small community in the northwestern Highlands, highlights how the opposition to these farms depended on the efforts of a handful of committed activists (Crowther *et al.* 2012). A significant part of this community was comprised of people who had moved to the area seeking an alternative lifestyle, and this factor was important in stimulating opposition to the development of fish farms. The campaign focused on 'aesthetic, ecological and livelihood arguments', although some opponents were also concerned about the impact of aquaculture on local fisher people (Crowther *et al.* 2012: 119). A key focus of contention in Scoraig was the role of the Crown Estate which, as mentioned above, owns the Scottish coastline (Phyne 1997). Activists argued that leasing agreements made by the Crown Estate should take damage to the seabed into account (Crowther *et al.* 2012). Some key campaigners identified their struggle as one of 'environmental justice', and they found support from Scottish environmental organizations and anti-salmon farm activists. But it was often difficult to generate broader community support, particularly given the Scottish government's vocal support for salmon aquaculture. In the end, the development in Scoraig did not go ahead, but researchers doubt that this outcome was a result of the campaign itself (Crowther *et al.* 2012).

State frameworks for property rights to inland and coastal waters are crucial to how struggles in Scotland and elsewhere develop. In general, property rights to marine environments have been much more complex than for land, with a variety of competing claims based on state ownership, private property and customary use continuing into the present, even in places where land-based commons were enclosed for purposes of capitalist accumulation several hundred years ago (Phyne 1997). Phyne characterizes such struggles around industrial-scale aquaculture as 'conflicts analogous to the struggle for enclosure' (1997: 74). As mentioned above, such struggles have continued to intensify as global actors push for 'rights based fisheries' – or the privatization of the ocean's commons – as a solution to the depletion of fish stocks around the world (World Forum of Fisher Peoples 2015).

Letting the salmon speak

Disputes over science are generally not at the centre of indigenous anti-salmon farming campaigns in Canada, but land and property are central although, as discussed below, in a different register. While the environmental impact of salmon farms is a major theme in the protests, concerns focus on how farming has transformed the nature of the fish as an entity and the labour relations and processes of production involved in its harvesting. As First Nations fishers moved from small quasi-independent producers to wage-labourers, the production of fish has been subsumed under the logic of capitalist production, fundamentally undermining indigenous livelihoods (Newell 1993; Menzies and Butler 2007; Menzies 2012). The announcement for the 1 March 2013 Vancouver demonstration that launched transnational action against salmon farms set out the reasons for action in stark terms:

> We have reached a tipping point and cannot stand idly by while our most important food is destroyed by the fish farming industry.
> Norwegian fish farms have colonized our waters and infected our wild salmon. Our salmon are forced to swim through the fish farm cesspools along their migratory routes. There have been findings [in salmon] of heart virus, salmon flu and brain tumors, all related to salmon farms, which have all been traced back to Norwegian sources. There is no safe passage for them! . . .
> For years British Columbians have tried to persuade this Norwegian industry to respect our Wild Fish, even offering to fund their transition to closed tanks, but they just keep expanding. 5000 people stood at the BC [British Columbia] legislature telling government to get salmon feedlots out of the ocean, away from our Wild Salmon, but they did not hear us.
> This is a time for people in Canada, Ireland, Scotland, Chile and wherever salmon farms are impacting wild salmon, to join forces and send a message

loud and clear to [the Canadian Department of Fisheries and Oceans] that wild salmon MUST be protected from the impacts of open-net salmon farms!!

Join the Voices for Wild Salmon to stand for the fish that brings life to our coast.[15]

At this and other similar events, indigenous activists with drums emblazoned with distinctive black and red stylized images of salmon were joined by environmental campaigners. Since that time, there have been related actions in many parts of British Columbia. For example, a 'Wild Salmon Caravan' started on 10 May 2015 and travelled huge distances across British Columbia – from Prince George in the north, along the Fraser River to the sea, including a protest at a planned site for a facility for processing hazardous waste in Chilliwack on 13 May,[16] – and culminated on 14 May in Vancouver.[17] Along the way, marchers gathered for prayers and offerings, songs and stories, performances and oral history, food and camping, hosted by the various coastal First Nations communities along the route.

Such protest actions combine First Nations symbols, songs and dances with artistic representations, flash mobs and social media campaigns. They bring together an alliance of different First Nations groups with environmentalists and concerned scientists, particularly Alexandra Morton, a marine biologist who has made identifying the dangers of salmon farming to marine ecosystems her life's work.[18] They reflect a synthesis of long-standing and contemporary activist tools that has been evident in the 'Idle No More' movement generally (Barker 2015). However, such perspectives on the threat of salmon farming to indigenous ways of life did not begin in 2012 – their long history is intertwined with the struggles of indigenous peoples against colonial regimes.

Indigenous traditions and practices around the salmon on the Canadian Pacific coast are distinctive in perpetuating into the present and future an alternative world that rejects the dichotomy between (wild) nature and (human) culture that is frequently observed in modernist conceptions of the world (Latour 2007). As one indigenous man said:

> Our tribe is known as the Salmon People . . . The traps, fish wheels, they all came from our people. . . . Sockeye [salmon] is our bread and butter. Sockeye is as important as the air we breathe . . . Salmon is as important to us as our language because that's who we are.
>
> (Ladd 2011: 362, quoting an interviewee from the Canadian Pacific coast)

Indigenous activists, anthropologists, archaeologists and others show how the concept of the 'salmon people' situates both people and fish as interdependent natives of the region, with the 'right to coexist with and fish for native salmon . . . as a cultural imperative that must be protected and handed down to future generations' (Ladd 2011: 361).

An examination of the archaeological record supports the idea that this indigenous vision of an interdependent and sustainable life with the salmon is not only possible, but has long been practised by the people of the Canadian Pacific coast and other peoples in the North Pacific (Campbell and Butler 2010; Colombi and Brooks 2012). A review of archaeological evidence dating back 4,000 years found that despite a population of significant size and very substantial harvests of salmon and other food resources, including for purposes of trade, the most important contributor to the sustainability of livelihoods over time was social practices and beliefs that restricted such harvests to levels that were sustainable, as well as reliance on a range of other food sources: '[T]he institutions, beliefs, and rituals known for the indigenous peoples of the Pacific Northwest had the effect of managing human behaviour so that salmon harvesting timing and intensity were moderated by some group or central decision-making process' (Campbell and Butler 2010: 21).

Similar complexes of practices were widespread across the region, suggesting that they were a response to common environmental and social conditions. They were embedded in a 'cultural repertoire' of activities that regulated fisheries and other sources of livelihood in ways that sustained access to these resources over time (Campbell and Butler 2010: 13). Yet these practices were not only a matter of conserving naturally occurring resources, but also involved humans adapting the natural environment to facilitate the interdependence of people and fish, such as through building stone fish traps and weirs (Menzies 2012). Thus in conjunction with the salmon itself, over thousands of years indigenous people in the region shaped the marine and riverine environments in order to facilitate their monitoring and sustainable harvesting of salmon (Menzies 2012). More recently, indigenous activists in some areas have revived these traditions by rebuilding traditional fishing weirs on rivers in the region, sometimes without seeking permission from the relevant authorities, echoing earlier kinds of civil disobedience such as the 'fish-ins' mentioned above. Such technologies are used not only for fishing purposes, but also serve to track the progress of the spawning fish, and thus combine surveillance of salmon numbers with traditional rituals and practices, such as the 'First Salmon Ceremony' before which fish may not be caught (Dale and Natcher 2014).

Given this long history of interdependence, the campaigns of First Nations people seeking 'justice' for the wild salmon are not directed at returning the fish to some pristine, pre-contact state (Menzies and Butler 2007; Menzies 2010, 2012). The indigenous view of the salmon is thus distinct from a conservation approach that aims to separate the salmon into 'nature' and protect it from human depredation. This is one example of how indigenous people evince a world in which cultural belonging, livelihood and environment are inseparable (Newell 1993; Schreiber and Newell 2006; Colombi and Brooks 2012; Cantzler 2015). This is a vision embedded in a temporality that brings together the past, present and future (Schreiber and Newell 2006). Stories, genealogies and rituals link the past to the present, while the maintenance of these practices, and the transmission of the narratives and rituals that sustain them, are seen as critical to their continuation into the future. At the centre of such world views is a relational respect for animal kin and taking only as much as required (Menzies 2013; Samson 2013), comparable to the persistence of indigenous values around food among people in the Chittagong Hill Tracts discussed by Ali and Vallianatos in Chapter 2. Again and again, indigenous activists speak of their stewardship of their lifeworld for the generations yet to come. As one First Nations member interviewed said: 'Sacrificing survival for short-term economic gain is not worth it' (Gerwing and McDaniels 2006: 268).

However, this is not just a matter of preserving indigenous 'culture.' As the example of the fishing weir mentioned above shows, indigenous ecological knowledge involves systematic practices such as enumeration of salmon numbers continued over time, as well as actual modification of the environment to facilitate interdependence. Such knowledge and practices were (and in some cases remain) embedded in social and moral systems that both transmit the knowledge and its methods of collection and also shape social practices to respond to that knowledge, such as limiting salmon fishing to certain rivers where stocks were perceived to be in need of restraint. This is not to say that such knowledge systems are outside modernity – they do not necessarily exclude new forms of knowledge, such as those generated by scientists (Menzies 2013). However, the continued dominance of scientific and expert forms of knowledge – as well as the lack of recognition of these as cultural forms (Wynne 1996) – mean that indigenous knowledge is rarely accorded full recognition as an alternative approach to addressing questions regarding natural resources. Despite some recognition of the need for decolonization of knowledge practices, in practice, such a perspective has hardly affected mainstream social or natural sciences (Menzies 2013).

Ironically, perhaps, even scientific accounts of the effects of human actions, including salmon farming, on the survival of the wild fish have sometimes noted how cultural representation of the salmon can be crucial to creating a sense of urgency around such endeavours (Campbell and Butler 2010). A species that in the past was 'elevated to the regal ranks of the tiger, the lion and the stag' has now 'fallen to the lowly status of mass-produced commodity ridden with disease and chemically dependent' (Coates 2006: 8, 98). As Campbell and Butler describe the task:

> We need to continue to support scientific studies, and find ways to improve habitats overall, but in the meantime, also put more resources into activities that promote development of social beliefs and traditions about the value of salmon and their ecosystems to our everyday lives.
>
> (Campbell and Butler 2010: 16)

Such a vision of the salmon is evident not only in the most recent campaigns against salmon farming, but also in a number of initiatives in recent years around the north Pacific that have sought to reinstate the fish as central to the people of the region (Coates 2006).

However, a purely cultural view of the centrality of the salmon risks eliding the historically-grounded specificity of indigenous claims, and their connection to long-standing struggles against colonial projects. In order to address the conditions in which many indigenous peoples find themselves, 'we cannot simply accept the way things are now as given. We must look at *how* certain unique features of indigenous peoples' way of life changed' (Samson 2009: 83, emphasis added). Mobilizing the power of different world views is a crucial move in the face of the global ocean grab, among other transnational struggles. Further research is needed on the potential of such transnational connections, whether to salmon, or to other ways of life. Rather than being an atavistic return to the past, the revival of indigenous practices aims at 'the creation of alternative futures that can build upon rather than erase their own inheritances', writes Samson (2013: 231).

'Rights' versus collective claims

In the European sites, a *collective* relational claim to the land has long disappeared, expunging a basis for social movements that remains potent for many indigenous communities. By contrast with a regime of at least nominally settled property rights, in the Canadian context, rights to land – and by extension to marine resources – are still in dispute, particularly in British Columbia where many treaty claims by indigenous groups have yet to be dealt with. The campaign against salmon farms along the Canadian Pacific coast is thus an aspect of broader struggles by indigenous people against colonialism. These are not postcolonial movements, but anti-colonial ones that are part of centuries-long efforts to realize indigenous self-determination within states that have sought to extinguish their aboriginal inhabitants and displace them from 'territories and livelihoods that traditionally brought them great wealth and prosperity' (Schreiber and Newell 2006: 218). As Barker writes, 'Canada is a settler colonial state, whose sovereignty and political economy is premised on the dispossession of Indigenous peoples and exploitation of their land base' (2015: 44). Many of the most egregious genocidal policies of the settler states have now been repudiated – most notably with apologies from the Australian and Canadian governments for past wrongs committed against their indigenous inhabitants. However, such apologies have not been followed by changes in the fundamental relations between aboriginal people and these states, and in particular, have not returned to them the resources they see as rightfully theirs (Schreiber and Newell 2006; Samson 2013). Elders of the Kwakwaka'wakw First Nation see salmon farming as 'a modern-day version of the cultural genocide' that began with the ban on the Potlatch ceremony that was central to the organization of the societies of what is now north western British Columbia (Schreiber and Newell 2006: 218).

Such collective visions of land and sea use are still under threat in Canada. In a parallel to how 'rights-based fisheries' are being used to enclose the oceans, 'rights' frameworks are also being deployed in land-claims negotiations with particular First Nations groups to cut off future claims and limit the size of indigenous traditional territories (Samson 2013; Samson and Cassell 2013). A major precipitating factor in the 'Idle No More' movement was proposals for sweeping changes to Canadian federal law relating to government relations with indigenous peoples and environmental protection (Barker 2015). While indigenous collective rights may be recognized at global level, within national spaces individualized rights and representation often serve to further fracture indigenous communities in the name of addressing pressing current concerns (Samson 2009, 2013). For example, factors such as the imposition of structures for representation of indigenous groups – most notably elected band councils – has diminished the role of traditional modes of collective deliberation and consensus formation (Schreiber and Newell 2006). These structures operate in order to fracture the collective will of indigenous groups and replace them with property relations more similar to those that pertain in Scotland and Ireland.

Yet although historically this has been its predominant role, the law has not only been used as a tool of dispossession. In Canada and Australia, recent court decisions recognizing that aboriginal titles to lands and waters have never been extinguished contributed to the upsurge in organizing around fisheries. In the wake of such cases, indigenous fishers in Canada and Australia are asserting rights to revive their conservation and harvesting practices in their historical sea and riverine territories (Menzies and Butler 2007; Dale and Natcher 2014; Cantzler 2015). Even in Norway, despite increasing recognition of the indigenous Sami people in Norwegian law and governance institutions – and Norway's reputation as a model for incorporating the rights of indigenous peoples into the political architecture of a liberal democratic state – fisheries and continuing traditional access to marine resources remain an area of tension (Broderstad 2014). A transnational perspective on struggles over salmon farms calls into question Norway's perceived accommodation of indigenous perspectives, given the role of Norwegian multinationals in the fish farming business and its impact on indigenous peoples beyond Norway.

Contrasts in the grounds of struggles over fish farming highlight differences in the social basis of these movements. While environmental movements tend to involve temporary alliances (Crowther *et al.* 2012), indigenous movements are enmeshed in long-standing kin-like relations attached to place. The former often involve developing a *new* awareness of environmental harms or risks, which may entail forming and mobilizing a community of people with a common experience of these environmental issues. By contrast, indigenous movements potentially draw on an existing community of fate based on collective attachment to place – formalized in indigenous land claims and treaty negotiations – with a shared mobilizing repertoire of stories, rituals and relationships (see Crossley 2002 on the importance of existing networks for the emergence and sustainability of social movements). Another contrast is in relation to the state: while environmental movements often seek state intervention to halt or mitigate environmental damage, indigenous actors frequently challenge the very legitimacy of the state agencies against which they contend – they situate themselves outside the realm of 'normal' politics. 'Instead of seeking inclusion within, or accommodation by, the broader society, Indigenous Peoples often demand rights to political self-determination and cultural autonomy' (Cantzler 2015: 75). Such mobilization is 'a uniquely cultural phenomenon' (Cantzler 2015: 75). Yet this 'culture' is deeply embedded in territory and eco-systems, as the survival of indigenous cultures are seen as inseparable from the places where they are practised (Samson 2009, 2013).

Narratives that entwine indigenous people with the past, present and future of the worlds they inhabit are central to such struggles.

Nature versus culture?

Highlighting how indigenous campaigns against fish farming assert alternative world views goes beyond pointing out the ways they deploy culture as a resource. The latter approach is implicit in much of the 'cultural turn' in social movement studies yet, as Jasper remarks, too often in this work, culture is seen as just another type of resource, and its moral and emotional dimensions, as well as broader embedding in a historically informed account of social and economic life is not fully theorized (Jasper 2010). As sketched out above, struggles over salmon farming in Scotland and Ireland tend to pit aquaculture advocates, proponents of 'local development' and governments against salmon sport fishermen, environmentalists and promoters of local tourism. But these struggles largely occur on the same ground – disputes over science, over economic benefit and over the impact of salmon farms on preferred *human* activities. They also occur in a present tense in which the current configuration of forces is seen as a given, rather than a historically contingent and contested terrain.

Conflicts between nature and culture are the focus of accounts of salmon farming using Actor Network Theory (Lien and Law 2011). As a combined theory and method, one aim of ANT can be to problematize the divide between a human social realm of 'culture' and the 'natural' world of material things, including non-human creatures (Latour 2005, 2007). Reflecting on his ethnographic research on salmon farms in Norway, in a conversation with Vicky Singleton about ANT, John Law talks of how, from the material practices and entanglements in and around the salmon pens, he is able to give a full account of the complex reality and the politics of salmon farming:

> So on the fish farm it is salmon all the way down. And people and government agencies and economic markets, and fish science and welfare concerns and local communities and feed and feedstock fish from Chile and big pharma, and food markets and chilled distribution chains and consumers.
>
> (Law and Singleton 2013: 493)

Studying these interconnections, an ANT account reveals the assemblages that actually make up a heterogeneous, interrelated and uncertain world that is emergent: 'Everything is related to everything else. And gets itself assembled, one way or another' (Law and Singleton 2013: 491).

The latest manifestation of ANT accounts is Marianne Lien's ethnography of salmon farms in Norway and Tasmania (Lien 2015). This deploys the metaphor of 'domestication' to explore its subject, while also questioning the narrative of 'progress' that lies behind this concept, revealing the contingent and uncertain character of efforts to domesticate the Atlantic salmon in industrial aquaculture. While Lien aims to 'tell the world differently' and 'cultivate our ethnographic imagination so that it [includes] the salmon' (Lien 2015: 168), her focus on the processes of raising salmon in farms and tanks seems a reductive way to achieve such aims. For instance, while she states that European animal welfare regimes have 'recently' made salmon 'sentient' (Lien 2015: 26), this view neglects how the fish has been always already a sentient being in indigenous perspectives. Law asserts:

> *To the extent that ANT explores the contingencies of power it also generates tools for undoing the inevitability of that power.* More strongly, it starts on the process of undoing its inevitability. For, at its best, ANT works on the assumption that other worlds are possible.
>
> (Law and Singleton 2013: 500–1, emphasis in original)

But these accounts neglect the other worlds that are *already here*, and are being undermined by the very existence of the farms they describe.

The kinds of assemblages produced around salmon farms are never neutral; conflicting world views and questions of justice are implicit within every network and assemblage. Flattening out the landscape of connections does nothing to high-light the vastly unequal dynamics of transnational power around aquaculture, and in starting from fish farms, ANT accounts can become a-historical and a-moral. Their focus evinces a romance with technology, reflected in an insistence on staying at the technical frontline of salmon farms, as Lien's (2015) book tends to do. Little or no account is provided of what is silenced or left out, and sticking to the here and now makes the historical background to how a particular network came to be formed largely disappear from view. As mentioned above, the perpetu-ation into the present of historical processes of dispossession and cultural sup-pression are central to how indigenous people experience phenomena such as salmon farms (Schreiber and Newell 2006; Samson 2009, 2013). While acknowl-edging that what to *describe* involves choice, as simplification will be necessary in any account, the ANT versions do not give an adequate account of the politics of these choices (Law and Singleton 2013; Lien 2015).

A brief examination of how powerful players in aquaculture seek to discredit opponents and alternative visions of the world highlights the deeply political nature of what comes into view in an account of this field. As powerful global actors, fish farming multinationals have developed sophisticated public relations responses to anti-salmon farm campaigns. For example, Mainstream Canada, a multinational involved in salmon farming in British Columbia, brought a defama-tion suit against the environmentalist Don Staniford and the Global Alliance Against Industrial Aquaculture (GAAIA) seeking a permanent injunction banning his claims that 'salmon farming kills'. The courts in British Columbia ruled in favour of the plaintiff, a decision upheld by Canada's Supreme Court. Staniford and GAAIA were ordered to pay damages and court costs estimated at CA$500,000 and were banned from 'further writing, printing, broadcasting, or causing to be written, printed, broadcast, or otherwise publishing' a set of fifty-two statements including, 'Warning: Salmon Farming can seriously damage the health of wild salmon' and 'salmon farming kills sea floor ecosystems'.[19]

Public relations efforts have been rolled out to dismiss scientific evidence on the impacts of farmed salmon on human health and the environmental effects of industrial aquaculture. Miller (2007) describes how a global network of pro-aquaculture organizations, websites and industry-supported scientists was mobilized in order to challenge a paper published in the rigorously peer-reviewed journal *Science* on high levels of organic chemicals called poly-chlorinated biphenyls (PCBs) found in farmed salmon produced in Scotland. Deceptive claims were made about the science in the paper, and industry advocates managed to convince journalists and policy makers to dismiss the findings. The Scottish government and the Crown Estate even supported the 'propaganda campaign' financially. Based on a detailed examination of records, many obtained through Freedom of Information requests, Miller concludes: '[V]ested interests operating together in a corporate-state two-step are able to manage science and silence critics – even where these emanate from the most prestigious scientific journals in the world' (2007: 68). He details how the aquaculture industry 'led journalists, policy makers and some sections of the public to believe that we were, in fact, victims of an orchestrated attack by environmentalists ... designed ... to destroy livelihoods and undermine healthy eating advice for ideological reasons' (2007: 68).

Such attempts to discredit scientific evidence about the effects of salmon farming on health and environment have been assisted by government agencies in Canada, Scotland and Ireland. As is noted more generally by some observers of disputes over salmon farms and fisheries, government agencies involved are responsible for conservation as well as development, and the latter priority usually wins out when these come into conflict (Phyne 1996; Broderstad 2014; World Forum of Fisher Peoples 2015). For example, the Cohen Commission noted serious conflicts of interest within the Canadian Department of Fisheries and Oceans, which has responsibilities both for promoting fisheries (including aquaculture) and protecting wild fish (Griggs 2012). Such conflicts are far from new; governments have been deeply involved in facilitating the global trade in salmon in its various incarnations for hundreds of years (Coates 2006).

Conclusion

So what difference does it make to call for 'justice' for the salmon in this global and highly unequal struggle, as the indigenous framing attempts to do? Both as a species and historically, the salmon is always already global, but this global character has been largely an aspect of forces that have enlisted the fish in colonial and capitalist projects (Coates 2006). By contrast, the effort to bring indigenous accounts of the salmon into a transnational campaign posits rights to a different world – not one conceived in the hegemonic terms of current debates about sovereignty, economic development, environment and so on. Although arguably some of the people in the communities contesting fish farms in Scotland and Ireland might also be considered 'indigenous', they have been repeatedly displaced and memory of any practices analogous to those in the Canadian Pacific coast has long disappeared.

As this brief account has attempted to show, the stories and rituals deployed by indigenous actors connect the past to the present and future – but also situate the salmon in an intimate relation with the human world. Such campaigns are one aspect of broader efforts to challenge the state-focused dominant order through indigenous efforts at revitalizing land- and sea-based practices (Samson 2013). Indigenous narratives require distinguishing between networks depending on their effects, incorporating a sense of history and taking a stand on political questions, something ANT approaches all too often fail to do, despite their pretentions to understand 'salmon all the way down'. 'Other worlds' are generally not made visible in the ANT accounts of the assemblages around salmon.

The alternative worlds visible in some indigenous anti-salmon farm campaigns are not only of consequence to those peoples and their ways of life. Soguk emphasizes the radical potential of indigenous peoples' defences of their different worlds, which constitutes a set of 'counter-spaces or "heterotopias" within modern orders' (2007: 3). These spaces of resistance represent a fundamental challenge to the 'epistemological hegemony of the modern territorial and nation-oriented imaginary' (Soguk 2007: 19), as they are places that have escaped this hegemony and its associated practices. As Wilson shows in Chapter 7 (p. 146), such resources can even be mobilized in places which have been 'wholly embedded in industrial capitalist food networks'. Research on such projects is inevitably place-based, so the question arises of how to generate fruitful connections across borders between sites for envisaging alternative futures. These projects are critical not only in rethinking the political as a space of deliberation, or relationships to territory and land, but also the very nature of being in the world, of livelihoods, of life itself.

Notes

1 From picture available at www.firstnations.de/fisheries/heiltsuk.htm (accessed 29 May 2015).
2 Mainstream media reports on these activities were rare, but independent media reported some of them, and other information is available on websites of various organizations involved. See 'Protest mot norsk lakseoppdrett' (Protest against Norwegian salmon farming), published 21 February 2012, available at www.nrk.no/sapmi/protest-mot-norsk-lakseoppdrett-1.10922125 (accessed 20 May 2015); 'Articles' published 28 February 2012, available at: www.fishnewseu.com/index.php?option=com_content&view=article&id=9951:fish-farm-protests-planned&catid=46:world&Itemid=56 (accessed 21 May 2015); and 'Idle No More – March for Wild Salmon', available at www.indigenousfoodsystems.org/content/idle-no-more-march-wild-salmon (accessed 21 May 2015).
3 There are three general categories of indigenous peoples in Canada (Metis, Inuit and First Nations) who are formally recognized under Canada's constitution as aboriginal. First Nation is the preferred term for aboriginal people formally recognized as Status Indians under the Indian Act of Canada.
4 Idle No More, Press Release, 10 January 2013, available at www.idlenomore.ca/about-us/press-releases/item/87- (accessed 21 May 2015).

5 In England and Ireland, rights over salmon were held by the aristocracy under royal prerogative (Coates 2006).
6 In the early nineteenth century, chemicals processed from the burning of kelp were an essential ingredient for the production of soap. The industry collapsed after cheaper substitutes were found.
7 Thus initial settlements of the contention over land in both countries in the 1870s and 1880s left out the landless, and struggles continued for many years after (Hunter 2000).
8 Version from a 1964 interview on salmon fishing in Loch Tay archived by the School of Scottish Studies, summary at www.tobarandualchais.co.uk/en/fullrecord/24880/7;jsessionid=900E33E29A82CD34EB37D3FEF00A7B1E (accessed 1 December 2015).
9 The most recent information is from the UN FAO at www.fao.org/fishery/culturedspecies/Salmo_salar/en (accessed 4 December 2015).
10 BC Environment Ministry statistics from 2012 put the value of local species farmed in the province that year at just over 4 per cent of the total sold (Drews 2013).
11 See www.fda.gov/ForConsumers/ConsumerUpdates/ucm472487.htm (accessed 4 December 2015).
12 For example, a study by marine biologists of the impact of salmon farms on the wild salmon population found that the presence of farms was a severe threat to certain wild fish species, particularly Atlantic salmon, sea trout and pink, chum and coho salmon. 'In many cases, these reductions in survival or abundance are greater than 50%,' the authors of this study write, arguing that, '[r]educing impacts of salmon farming on wild salmon should be a high priority' (Ford and Myers 2008: 0411–0412).
13 Sockeye is a particular species of the *genus* salmon that is found around the Pacific Rim, and is particularly prized by sport and aboriginal fishers (Coates 2006).
14 See for example, www.educationscotland.gov.uk/scotlandsstories/finnmaccool andthesalmonofknowledge/index.asp, www.celtic.org/salmonofknowledge.html and http://treesforlife.org.uk/forest/mythology-folklore/salmon/ (all accessed 14 July 2015).
15 'Idle No More – March for Wild Salmon', (undated) available at: www.indigenous foodsystems.org/content/idle-no-more-march-wild-salmon (accessed 20 May 2015).
16 On the proposed hazardous waste facility, and opposition to it by the local First Nations group, see www.theglobeandmail.com/news/british-columbia/first-nation-wants-say-on-proposed-hazardous-waste-site-near-fraser-river/article22825775/ (accessed 28 April 2015).
17 See tinyurl.com/WildSalmonCaravan (accessed 28 April 2015).
18 See, for example, Morton's film 'Salmon Confidential', available at www. salmonconfidential.ca/, and her blog, http://alexandramorton.typepad.com/alexandra_morton/ (both accessed 1 November 2015).
19 These are excerpts from the courts' judgments, published on the website: http://salmonfarmingkills.com/lawsuit (accessed 3 June 2015). All fifty-two banned statements are listed on this website.

References

Barker, Adam J. (2015) ' "A direct act of resurgence, a direct act of sovereignty": Reflections on Idle No More, Indigenous Activism, and Canadian Settler Colonialism'. *Globalizations* 12 (1): 43–65.

Brattland, Camilla and Dorothee Schreiber (2011) 'Salmon voices: Indigenous peoples and the fish farming industry, report on October 2011 conference'. Rachel Carson Center. www.carsoncenter.uni-muenchen.de/download/events/conference_reports/111007_salmon_voices_confrep.pdf (accessed 24 April 2015).

Broderstad, Else Grete (2014) 'Implementing indigenous self-determination: The case of the Sámi in Norway'. In Marc Woons and Ku Leuven (eds) *Restoring Indigenous*

Self-Determination. Bristol, UK: E-International Relations, 80–7. www.e-ir.info/wp-content/uploads/2014/05/Restoring-Indigenous-Self-Determination-E-IR.pdf (accessed 24 April 2015).

Bryden, John and Ian Scott (1990) 'The Celtic fringe: State sponsored versus indigenous local development initiatives in Scotland and Ireland'. In Walter B. Stöhr (ed.) *Global Challenge and Local Response: Initiatives for Economic Regeneration in Contemporary Europe*. London and New York: United Nations University Press, 90–132.

Cahill, Ann (2014) 'Full-scale war over Galway fish farm'. *Irish Examiner*, 26 February. www.irishexaminer.com/ireland/full-scale-war-over-galway-fish-farm-260075.html (accessed 30 April 2015).

Campbell, Sarah K. and Virginia L. Butler (2010) 'Archaeological evidence for resilience of Pacific Northwest salmon populations and the socioecological system over the last ~7,500 years'. *Ecology & Society* 15 (1): 1–20.

Cantzler, Julia Miller (2015) 'The translation of indigenous agency and innovation into political and cultural power: The case of indigenous fishing rights in Australia'. *Interface: A Journal For and About Social Movements*, 5 (1): 69–101. www.interfacejournal.net/wordpress/wp-content/uploads/2013/05/Interface-5-1-Cantzler.pdf (accessed 4 March 2015).

Coates, Peter (2006) *Salmon*. London: Reaktion Books.

Colombi, Benedict J. and James R. Brooks (eds) (2012) *Keystone Nations: Indigenous Peoples and Salmon across the North Pacific*. Santa Fe, NM: School for Advanced Research Press.

Crossley, Nick (2002) *Making Sense of Social Movements*. Buckingham, UK and Philadelphia, PA: Open University Press.

Crowther, Jim, Akiko Hemmi and Eurig Scandrett (2012) 'Learning environmental justice and adult education in a Scottish community campaign against fish farming'. *Local Environment* 17 (1): 115–30.

Dale, Chelsea and David C. Natcher (2015) 'What is old is new again: The reintroduction of indigenous fishing technologies in British Columbia'. *Local Environment* 20 (11): 1309–21.

Drews, Keven (2013) 'Fish farm commits to Atlantic salmon in B.C.' *The Huffington Post*, 30 November. www.huffingtonpost.ca/2013/11/30/bc-fish-farms-atlantic-salmon-cermaq-aquaculture_n_4365263.html (accessed 30 April 2015).

Ford, Jennifer S. and Ransom A. Myers (2008) 'A global assessment of salmon aquaculture impacts on wild salmonids'. *PLoS Biol* 6 (2): e33.

Gerwing, Kira and Timothy McDaniels (2006) 'Listening to the salmon people: Coastal First Nations' objectives regarding salmon aquaculture in British Columbia'. *Society and Natural Resources* 19 (3): 259–73.

Gillis, Damian (2015) 'Bad boy salmon activists teaming up in Norway'. *The Common Sense Canadian*, 3 February. http://commonsensecanadian.ca/bad-boy-salmon-activists-teaming-up-in-norway-staniford-oddekalv-farm/ (accessed 1 May 2015).

Griggs, Ray (2012) 'The recommendations of the Cohen Commission report'. *The Common Sense Canadian*, 18 November. http://commonsensecanadian.ca/the-recommendations-of-the-cohen-commission-report/ (accessed 1 May 2015).

Gross, Liza (2008) 'Can farmed and wild salmon coexist?' *PLoS Biol* 6 (2): e46.

Hunter, James (2000) *The Making of the Crofting Community* (2nd revised edn). Edinburgh: John Donald Publishers Ltd.

Jasper, James M. (2010) 'Social movement theory today: Toward a theory of action? Social movement theory today'. *Sociology Compass* 4 (11): 965–76.

Ladd, A. E. (2011) 'Feedlots of the sea: Movement frames and activist claims in the protest over salmon farming in the Pacific Northwest'. *Humanity and Society* 35 (4): 343–75.

Latour, Bruno (2005) *Reassembling the Social: An Introduction to Actor-Network-Theory*. Oxford and New York: Oxford University Press.

——(2007) 'We have never been modern'. In Craig J. Calhoun (ed.) *Contemporary Sociological Theory* (2nd edn). Oxford and Malden, MA: Blackwell, 448–60.

Law, John and Vicky Singleton (2013) 'ANT and politics: Working in and on the world'. *Qualitative Sociology* 36 (4): 485–502.

Lien, Marianne Elisabeth (2015) *Becoming Salmon: Aquaculture and the Domestication of a Fish*. Berkeley, CA: University of California Press.

Lien, Marianne Elisabeth and John Law (2011) '"Emergent aliens": On salmon, nature, and their enactment'. *Ethnos* 76 (1): 65–87.

Menzies, Charles R. (2010) 'Dm sibilhaa'nm da laxyuubm Gitxaała: Picking abalone in Gitxaała territory'. *Human Organization* 69 (3): 213–20.

——(2012) 'The disturbed environment: The indigenous cultivation of salmon'. In Benedict J. Colombi and James R. Brooks (eds) *Keystone Nations: Indigenous Peoples and Salmon Across the North Pacific*. Santa Fe, NM: School for Advanced Research Press, 161–82.

——(2013) 'Standing on the shore with Saaban: An anthropological rapprochement with an indigenous intellectual tradition'. *Collaborative Anthropologies* 6 (1): 171–99.

Menzies, Charles R. and Caroline F. Butler (2007) 'Returning to selective fishing through indigenous fisheries knowledge: The example of K'moda, Gitxaała territory'. *The American Indian Quarterly* 31 (3): 441–64.

Miller, David (2007) 'Spinning farmed salmon'. In David Miller and William Dinan (ed.) *Thinker, Faker, Spinner, Spy: Corporate PR and the Assault on Democracy*. London: Pluto Press, 67–93.

Newell, Dianne (1993) *Tangled Webs of History: Indians and the Law in Canada's Pacific Coast Fisheries*. Toronto: University of Toronto Press.

Page, Justin (2007) 'Salmon farming in First Nations' territories: A case of environmental injustice on Canada's west coast'. *Local Environment* 12 (6): 613–26.

Phyne, John (1996) 'Biological warfare: Salmon farming, angling tourism and the sea trout dispute in the west of Ireland, 1989–1995'. *Irish Journal of Sociology* 6 (1): 1–24.

——(1997) 'Capitalist aquaculture and the quest for marine tenure in Scotland and Ireland'. *Studies in Political Economy* 52 (1): 73–109.

Samson, Colin (2009) 'Indigenous peoples' rights: Anthropology and the right to culture'. In Rhiannon Morgan and Bryan S. Turner (eds) *Interpreting Human Rights: Social Science Perspectives*. London and New York: Routledge, 68–86.

——(2013) *A World You Do Not Know: Settler Societies, Indigenous Peoples and the Attack on Cultural Diversity*. London: Human Rights Consortium, Institute of Commonwealth Studies.

Samson, Colin and Elizabeth Cassell (2013) 'The long reach of frontier justice: Canadian land claims "negotiation" strategies as human rights violations'. *The International Journal of Human Rights* 17 (1): 35–55.

Schreiber, Dorothee and Dianne Newell (2006) 'Why spend a lot of time dwelling on the past? Understanding resistance to contemporary salmon farming in Kwakwaka'wakw territory'. In Arif Dirlik (ed.) *Pedagogies of the Global: Knowledge in the Human Interest*. Boulder, CO and London: Paradigm Publishers, 217–32.

Seafood Choices Alliance (2005) 'Salmon aquaculture'. *It's All about Salmon.* http://seafoodchoices.com/resources/afishianado_pdfs/Salmon_Spring05.pdf (accessed 9 April 2015.

Soguk, Nevzat (2007) 'Indigenous peoples and radical futures in global politics'. *New Political Science* 29 (1): 1–22.

Wilkes, R. (2006) 'The protest actions of indigenous peoples: A Canadian–U.S. comparison of social movement emergence'. *American Behavioral Scientist* 50 (4): 510–25.

World Forum of Fisher Peoples (2015) 'A New Report on Ocean Grabbing'. http://worldfishers.org/2014/08/21/new-report-ocean-grabbing/ (accessed 24 April 2015).

Wynne, Brian (1996) 'May the sheep safely graze? A reflexive view of the expert–lay knowledge divide'. In Scott Lash, Bronislaw Szerszynskis and Brian Wynne (eds) *Risk, Environment and Modernity: Towards a New Ecology.* London: Sage, 44–83.

4 Food sovereignty, permaculture and the postcolonial politics of knowledge in El Salvador

Naomi Millner

In this chapter I focus upon the development of agroecological knowledge systems in El Salvador over the last twenty years, with emphasis on their capacity to generate and sustain postcolonial food networks. Agroecology is the study of ecological processes that operate in agricultural production systems. First used in the early twentieth century, the term was popularized after the Second World War when increasing awareness of the environmental consequences of industrializing agri-food technologies led to the emergence of new networks and forums of knowledge production that emphasized a systems perspective. In Central America such networks emerged through widespread critiques of Green Revolution technologies introduced during the 1960s, and were consolidated through the development of a farmer-to-farmer (*campesino a campesino*; CaC) model for testing and sharing traditional agricultural techniques. The CaC model of organizing agricultural production was mobilized on a significant scale in El Salvador during the 1980s, when the country was reeling from a twelve-year civil war. The model was adapted for the Salvadorian context, and has also been hybridized with external forms of organizing associated with conflict-resolution and international aid initiatives.

The analysis in this chapter relates to the development of such alternative food networks (AFNs) that emerged in multiple locations in El Salvador during the 1990s, as associated with the practices of agroecology and, more specifically, permaculture. Permaculture is an agroecological approach to food production that employs a systems perspective by focusing its interventions on the points of interconnection between qualitatively diverse systems of biophysical, socio-economic and socio-cultural life. Permaculture is also a principle of environmental design that strives to establish 'permanent agricultures' or 'permanent cultures' – multi-species ecologies that support their various human and non-human components through mutually enhancing feedback loops and interactions. As in agroecology, traditional and indigenous agricultural techniques are strongly valued in permaculture, although these are always tested experimentally against existing practices. Permaculture as a term is less widespread than agroecology, although in contexts like El Salvador the two terms are highly complementary.

In this chapter I argue that the example of permaculture in El Salvador – particularly the CaC model for production and knowledge exchange and creation

– illustrates key conditions for the emergence of postcolonial AFNs. AFNs emerging in a post-conflict context are not automatically at odds with dominant knowledges and power relations associated with colonial histories. However, the legacies of popular education in El Salvador are being appropriated through agroecological networks like those configured through permaculture to critique and rework conventional principles of food security and development. Through pedagogical practices that legitimize *campesino* and indigenous experiences, 'tradition' is reimagined, not as a relic of a backwards past, but as a lively site of knowledge and expertise. As a consequence, indigenous and small-scale practices are being shared across international borders and consolidated into vibrant agricultural solutions. Within such transnational movements, indigenous tropes like *Terra Madre* (Mother Nature) are mobilized to express shared commitments, without presuming 'sameness' between diversely-situated groups. My argument consequently hinges upon a definition of a postcolonial politics of knowledge that underpins AFNs in El Salvador and, by implication, other contexts in the global South.

To make this argument I draw on ethnographic fieldwork and data produced within participative workshops held over two periods from December 2012 to January 2013 and March to April 2014, and in two main sites. Suchitoto, a small colonial town in the region of Cuscatlán, was an important guerrilla hub during the civil war and today hosts a central permaculture demonstration site (see Figure 4.1). Torola is a smaller town in the rural region of Morazán, and was

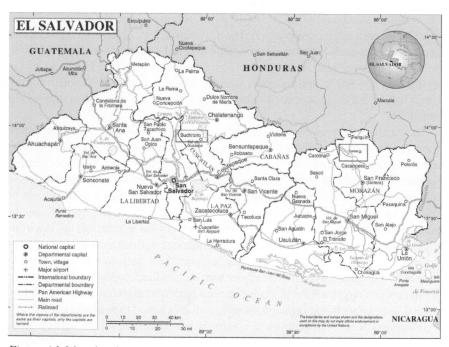

Figure 4.1 Map showing research sites (Suchitoto and Torola).

Source: https://en/wikipedia.org/wiki/list_of_cities_in_el_salvador#/media/File:Un-el-salvador.png.

also an important site of guerrilla activities during the war. The regions surrounding these two towns host the most active permaculture networks, although permaculture and agroecological activities exist elsewhere. This chapter draws primarily on data collected during the second research visit (March to April 2014) during which I carried out eight four-hour participatory workshops, co-designed with my partners, and thirty-two interviews with small-scale farmers, local, national and international non-governmental organization (NGO) representatives and regional and municipal governors. After getting the appropriate consents, audio and video capture of interviews and fieldwork enabled a second translation of my data upon my return to the United Kingdom. Combining methodologies helped to illuminate the way that concepts such as food sovereignty were perceived by different actors and provided multiple opportunities for participants to contribute towards, and correct, my narratives of the movement. In the following sections I unpack material on AFNs and food sovereignty in relation to postcoloniality before providing detail of the specific histories and geographies of the emergence of AFNs in El Salvador. I then discuss criteria for evaluating the postcolonial dimension of food alternatives within the context of permaculture in El Salvador, and conclude with reflections on key ways that the permaculture example helps us assess a postcolonial politics of knowledge and tradition.

AFNs, food sovereignty and the postcolonial critique

Alternative food networks have assembled against a backdrop of growing dissatisfaction with globalizing food regimes, focusing on the environmentally destructive effects of large-scale agriculture, the nutritional narrowing associated with export-oriented monocultural crops and the heavy use of chemicals associated with health issues and soil degradation. In recent years those seeking alternatives to industrialized food systems have looked beyond earlier concerns with organic production to a wide range of food production, distribution and retail activities presented as alternatives to conventional food systems. These include farmers' markets and farm shops (Hinrichs 2003), direct marketing schemes and Community Supported Agriculture projects (Hinrichs 2007), vegetable box schemes (Purdue *et al.* 1997), community gardens and food cooperatives. Literatures focusing on growing AFNs have noted the revalorization of short food supply chains, whilst emphasizing the way that shorter chains are also reliant on new technologies and forms of connectivity made possible by globalizing cultures (Harris 2010). As such, AFNs are positioned by scholars as specific responses to processes of global economic restructuring, rural decline and redevelopment, environmental concerns and progressive political ideals (Whatmore *et al.* 2003; DuPuis and Goodman 2005; Jarosz 2008). Embedded within such multi-scalar processes, AFNs seek to localize food systems and encourage contact between food producers and consumers, working to re-spatialize food systems that are seen to have become 'placeless'.

Though such work has pointed towards helpful synergies between diversely situated networks, important scholarly work remains to be done in conceptualizing such alternatives in postcolonial terms, not least because the empirical examples mobilized within the AFNs' literature tend to rely heavily on the experiences of highly industrialized countries – as emphasized in the introduction to this volume. Moreover, as has been convincingly raised by a number of critical essays on alterity within AFNs, such networks also rely on new kinds of globalizing processes that do not necessarily imply an equality of access or participation (Leyshon *et al.* 2003). Rather than writing uncritically about 'globalizing alternatives', we need to remain aware of the complex and disparate ways that alternative practices are assembled at particular sites, always with variation, and always with different configurations of spatiality, power and knowledge relationships. Just as scholars have demonstrated the way that the 'travel' of neoliberal policies results in a highly uneven neoliberal regime (Leyshon *et al.* 2003) so we must acknowledge specific geographical histories and broader networks of association in order to give a full account of the geographies of alternative food practices.

I argue that the critique of the uneven power and knowledge relations within AFNs finds a parallel in the recent attention to food sovereignty among different publics and the scholarship that is responding to their claims. Indeed, the appearance of food sovereignty as a fast-growing transnational social movement and political claim has much to offer studies of AFNs, particularly in relation to postcoloniality. As authors in this volume suggest (in Chapters 3, 6 and 7), these concepts have not yet been brought into dialogue. Whilst the term 'food sovereignty' was coined during the 1980s (Edelman 1998), its reverberation through diverse geographical contexts has been linked with the emergence of the social movement *La Via Campesina* (LVC), or 'The Peasant's Way', during the early 1990s, against a backdrop of global agrarian crisis, exacerbated by withdrawal of support for domestic agricultural sectors across the global South (Edelman 2005). Food sovereignty movements are producer-led, and therefore do not necessarily entail networks of food processing and distribution; nevertheless, the concept specifically addresses long production–consumption chains and uneven divisions of labour, and has been highly active in developing alternatives through experimentation with local economies. Moreover, the knowledge politics worked out within the food sovereignty movement has much to offer the notion of postcolonial AFNs since both actively work to destabilize historical forms of privilege associated with the production of knowledge and organization of roles within society.

The historical emergence of food sovereignty has also coincided with the development of large-scale, bottom-up critiques of postcolonial efforts to address unequal access to food. Founding members of LVC were unified by dissatisfaction with 'food security' terminology as it had been elaborated in United Nations forums since 1974; in particular they challenged assumptions that global hunger could be addressed through market-based solutions (McMichael 2014). LVC formally introduced food sovereignty at the United Nations World Food Summit of 1996 as a kind of 'agrarian citizenship' (Wittman 2009) requiring correlative rights for small-scale farmers. It was maintained that communities, including national communities, would be better equipped to solve issues of long-term ecological unsustainability if they could cultivate a full range of food crops within their borders, rather than relying on export crops. The vocabularies of human rights and sustainable development have since been reworked through food sovereignty manifestos, such as the seminal Nyéléni Declaration written collectively in 2007, and today the movement boasts a 'peasant internationalism' comprising of hundreds of thousands of small-scale producers (Martinez-Torres and Rosset 2010).

The recent food sovereignty literature provides an important corrective to the geographical bias of AFN studies to date, emphasizing the experiences of small-scale farmers in the global South, and attending to an important postcolonial voice emerging from outside the academy. As emphasized earlier, much of the literature on AFNs to date has focused on examples within the global North (although see for example Freidberg and Goldstein 2011; and Goodman 2004) and can be accused of reproducing problematic ontological assumptions associated with the longer history of modern Western thought. More generally, the agri-food literature is accused of maintaining a dualistic separation between 'nature' and 'culture' within its analyses of eco-social production and food consumption (Goodman 1999) – divisions which simply do not exist within many indigenous systems of knowledge. The tendency to regard land managed by indigenous peoples as 'empty space' has been called an ontological act of dispossession, for its wilful disregard of existing forms of inhabiting and knowing environments, which enables the second, material act of dispossession, in the form of enclosure (Makki 2014). This knowledge politics is foregrounded within food sovereignty campaigns and networks, which also emphasize the cultural loss associated with industrial agriculture – for example, where rapid industrialization leads to the erosion of long-standing soil and seed conservation practices within single generations.

An alternative, postcolonial knowledge practice has been experimentally developed within LVC through its own processes of decision-making and organizing. Central to the food sovereignty movement are forums of dialogic knowledge exchange (*diálogo de saberes*), which foreground diverse forms of traditional knowledge and expertise, with full recognition of the non-compatibility of these systems to each other (Pimbert 2006; Martinez-Torres and Rosset 2010). This knowledge politics is supported by the growth and development of agroecological networks of production and exchange, which emphasize the specificity of particular agricultural solutions to particular climatic and cultural contexts, and the importance of experiential knowledge in relation to such specificities. There is a pervasive assumption throughout such networks that 'bottom-up' social assemblages are consequently the appropriate vehicle for managing the complex interrelation of food production and environmental conservation practices. Central to such accounts is a retheorization of the centres and peripheries of globalizing agricultural systems (Fairbairn 2014), and a fresh emphasis on the *politics of knowledge* at play. Drawing on postcolonial theory and understandings of democratic knowledge production, the emphasis here is on mobilizing bottom-up appropriations of self-proclaimed centres of knowledge, and legitimizing the expertise and 'solutions' that are already being developed elsewhere. In the case of Salvadorian permaculture, the development of ecological solutions also involves a critique and destabilization of the way that conventional agricultural management systems concentrate hierarchies of knowledge. In permaculture, *campesinos* (peasant farmers) are the experts and experimenters that state technicians need to learn from, rather than the 'poor' that technicians instruct and administrate.

The appearance of claims to food justice and food sovereignty thus also speak to long-established social concerns with how to write, represent or attend to excluded social voices. Within postcolonial studies and political ecology, research strategies have, particularly since the 1980s, been influenced by historical movements in the global South such as the Subaltern Studies group, a primarily South Asian group of scholars who emphasized the rewriting of histories from non-élite or bottom-up perspectives. More recently, social scientists have revisited their relation to such movements, unsettling notions of 'speaking for the subaltern', which Spivak (1985) argues, keeps repressed groups in a silenced position. In contexts of both the global South and North, fresh emphasis has been placed on tracing discourses of resistance and the claims of insurgent networks (Guha 1999), together with the hierarchical structures of knowledge that perpetually reframe them. This also involves deconstructing problematic oppositions between indigenous knowledge and science (Agrawal 1995), and 'provincialising' (Chakrabarty 2009) social theories which have emerged in tandem with cultural imperialism. Such claims also call for a shift in emphasis from the academic production of 'political speech' to 'political listening'. To take situated practice seriously as a claim to knowledge is also to promote the decolonization of scholarly discourse through attention to 'parts' that have conventionally been excluded or disregarded (Heynen 2010). In the context of alternative food regimes this means exploring the role of *in situ* forms of knowledge – such as that associated with the conservation of seeds and soils – in constituting alternatives, and in enabling the translation of alternatives from one site to another. It is likely, of course, that there are aspects of agro-ecological practice which cannot be easily translated or scaled-up, so it is important to be clear about what it would mean to broaden and deepen a democratic politics of knowledge. This calls for an ethics of translation which acknowledges what cannot be translated, and what ought not to be translated.

Recently, anthropologists such as Philippe Descola (2010) and Eduardo Viveiros de Castro (2004) have shown how the idea of democratizing knowledge shifts when we acknowledge that the process of translation also connects different 'worlds' of inhabitation, or ontologies of nature. Within cultures that share few historical pathways in common with Western science, agricultural expertise and knowledge of soil or seeds are characterized by very different ways of knowing environments. Indeed, the idea that culture and nature *can* be separated to describe the domains of human creation and non-human life is a highly specific Western production. This insight is very important to the analysis of the research context in question, as new agroecological and permaculture networks can be seen to embody new techniques for translation between *in situ* and traditional practices and human/non-human relations. In the example of organic compost production discussed below, indigenous practices linked with specific properties of tropical forest components and their dynamic properties are brought into dialogue with experimental practices from other parts of the world. Permaculture works to conserve and mobilize site-specific indigenous practices, as well as the implicit principles of relationality these rely upon. On the other hand, permaculture itself acts to perform its own ethics of translation, by testing these principles in turn, and discarding traditional practices that do not work in favour of other practices that are more effective. To be postcolonial, in this sense, is not to be conservative.

In terms of decolonizing and pluralizing alternatives in this way, food sovereignty has been heralded as a potential model for knowledge production, because of the way it has yielded 'translatable' shared claims whilst also respecting highly specific cultures of practice (Pimbert 2006). Martinez-Torres and Rosset (2010) link this distinctiveness with the *dialogo de saberes* practice developed by LVC: the concept of a dialogue between different kinds of knowledge and ways of knowing which precedes deliberation and planning. For such scholars, the food sovereignty framework is so powerful precisely because it emerges from shared experiences of dispossession and colonialism in diverse contexts, elaborated into a shared discourse through platforms including early agroecological movements and peasant training schools throughout the Americas, Africa and Asia. Much of what is distinctive about this movement emerges from the way this shared language of experience allows for the negotiation of internal differences – disagreement over the basic politic unit (family, community or collective) or the appropriate vehicle for agency (workers, families or militants), for example – without collapsing them into agreement (Martinez-Torres and Rosset 2010). Such networks and their insurgent claims have been called 'pre-figurative' institutions: collective bodies which practise into being the kind of democratic knowledge politics they consider absent from global forums of decision-making. Pre-figuration entails bottom-up critiques of globalizing political economies but also the embodiment of future alternatives and forms of inhabitation. From such perspectives, movements like food sovereignty not only make claims on policy domains, but actively seek new ways of living.

AFNs in Central America

Food and colonial violence in El Salvador

The Salvadorian permaculture movement intersects in important ways with articulations of food sovereignty, although it operates at a remove from the activities of LVC. To understand the way in which AFNs have been assembled in El Salvador in relation to a postcolonial politics, it is important to understand the ways in which globalizing food regimes in this country have been inflected by highly specific state governance regimes, civil conflict and social movements responding to colonial violence in different forms. Co-operative forms of small-scale production and broader networks like permaculture evolved in El Salvador in response to oppressive oligarchic regimes, which saw the military allied with a small number of land-owning families, themselves privileged by historical appropriations of both land and wealth.

For Hume (2009) and Montobbio (1999), dependency on coffee provides the key link for understanding resistance, land politics and colonial history in El Salvador. Whilst gaining its independence from Spanish colonial rule in 1821, state governance was initially concentrated in the hands of a small number of powerful families who maintained their élite status through a series of constitutional amendments, leading towards the liberalization of trade from the 1840s, when coffee replaced indigo as a key export crop. Liberal agitations during the 1880s culminated in a coffee revolution, when new lands were opened to cultivation through government-backed appropriations of indigenous communally-owned lands. The powerful families of El Salvador converted their land for the production of coffee, supported by government subsidies, forming a new oligarchy in association with the military. Within this regime, the *campesino* minority and indigenous peoples were progressively impoverished and displaced. During the colonial era the Spanish had been mostly supportive of the concept of Indian communal lands, and indigenous peoples had maintained their livelihoods through share-cropping and subsistence growing. However, the large expanse of indigenous communal lands placed commercial and governmental élites into competition for land and labour. From 1882 private ownership was declared the only recognized form of land ownership, communal rights were legally marginalized, and it became increasingly difficult for indigenous peoples to retain communal landholdings. In 1912 the government formed a National Guard, using Spain's *Guardia Civil* as a model, to watch over coffee-growers' interests in the face of growing rebellions. Most notable was the suppression of 1932 Salvadorian peasant massacre (*la Matanza*), during which between 15,000 and 30,000 indigenous peoples, as well as thousands of *campesinos*, were killed by state forces. During this period, indigenous populations virtually disappeared from El Salvador.

After the Second World War, traditional patron–client relationships on estates were transformed by highly coercive wage labour relations as cattle-raising, cotton cultivation and sugar expanded El Salvador's trade repertoire. However, the coffee élite, backed by the military, used such changes to further reinforce their dominance (Montes and Gaibrois 1979). Huge migrations across the country took place during the harvest months, while many sought work in distant labour markets. To deal with the rising dissent, rural unions were made illegal and coffee estates (*haciendas*) were made increasingly secure. There were sporadic attempts to reform land tenure and labour relations, but the core alliance was defeated in 1944, 1960 and 1972. The best agricultural land was concentrated among a small group of coffee plantation owners, while the high labour demand was met by a permanent, unfree labour force living in highly restrictive conditions. Debt peonage was the norm, elections were tightly controlled, and in some places sterilization was practised as an imposed form of birth-control (Roseberry 1991).

It was during this period that Green Revolution technologies were also experimentally introduced in El Salvador. Following discussions with the US Vice-President Henry Wallace, from 1943 the Rockefeller Foundation and the Consultative Group for International Agricultural Research (CGIAR) made the Mexican Agricultural Program its key 'developmental' intervention in the region. With the remit of intensifying food production in impoverished areas, the CGIAR was also involved from the outset with the collection of indigenous germplasm for commercial experimentations, drawing comparisons with the 'plant-hunters'[1] of the 1600s (Mangelsdorf 1951). Over the next eight years (from 1943), projects based on the Mexican model were implemented throughout Central America and Brazil, mostly under the auspices of the US Department of Agriculture. The CGIAR and Rockefeller Foundation persuaded national governments that failure to introduce 'modern' solutions would lead 'underdeveloped' countries to accept communist promises and systems (Carey 2009). A number of Central American governments consequently adopted recommendations primarily based on national security fears. The economic benefits of such schemes would have been minimal, since the new agricultural supplies were mainly obtained through US-based companies. Meanwhile, the 'miracle' high-yielding varieties required higher and higher inputs of fertilizers as soils became impoverished.

Violence deepened further within El Salvador as labour conditions worsened and agrarian autonomies deteriorated. In 1976 coalitions of landlords and military hardliners brutally derailed a reformist government's attempt at a limited agrarian reform along the coastal plains, while strikes for higher wages in 1979 were repressed with increasing violence. A coup on 15 October 1979 led to the killing of anti-coup protesters by the government as well as anti-disorder protesters by the guerillas, and is widely regarded as the tipping point from such uprisings towards civil war. In 1980 the Farabundo Martí National Liberation Front (FMLN) formed, as a coalition of the five principal left-wing guerilla groups, and immediately announced plans for an insurrection against the government, which began on 10 January with the FMLN's first major attack. The civil war lasted until 1992, with shifting loyalties overseas largely reflecting global sympathies and antipathies towards communist Russia.

Agroecological networks

Widespread popular education networks multiplied in El Salvador at the end of the 1970s in association with the guerrilla uprisings, and through these networks, Green Revolution technologies became connected in the popular imagination with land inequalities. Perhaps surprisingly, just a decade beforehand, few of the networks or resources for revolt were in place. A survey of *campesino* culture conducted in 1973 by the Jesuit sociologist Segundo Montes found that the rural poor in El Salvador were fatally resigned to poverty, with low social solidarity and high competition for land and jobs (Montes and Gaibrois 1979). Wood (2003) concluded that two factors were crucial in bringing about resistance: liberation theology and the organizing practices of tiny guerrilla organizations that built upon such practices. The liberation theology movement, reaching El Salvador from other parts of Latin America, entailed the rejection of established authoritarian readings of the Bible in favour of a 'preferential option for the poor' (Smith 1991: 24). Previously, authority figures of the Church and State had been associated with the divine will of God. The idea that the oppression of the poor was an obstacle to God's will created a moral impetus for organized resistance, to which many became committed to the point of losing their lives (Pearce 1986). Sparked by calls for revival at the Vatican from 1962 to 1964, Carlos Rafael Caburrús (1983: 135), a Guatemalan Jesuit, described the movement as an 'unblocking' of radical *campesino* fatalism across Latin America.

From the late 1960s radical Catholic priests inspired by liberation theology began to denounce the Salvadorian government from the pulpit and to publicly record the abusive harvesting practices of local landlords. Across El Salvador a peripatetic network of priests and other active intermediaries, known as catechists, began to provide pastoral support for covert Bible study groups in parishes (*cantones*) across El Salvador. The study groups drew strongly on the pedagogical principles established by the Brazilian popular educator, Paulo Freire, for his adult literacy programmes in the 1970s, which were premised on the capacity of individuals and groups to make their own interpretations of their situation. Whilst framed more in terms of cultural oppression and Marxist ideas of the owner- ship of production, Freire's ideas provided the basis for postcolonial critiques of regimes throughout Latin America, legitimizing *campesino* readings of power relations and articulations of liberation. The ensuing feeling of equality was critical to subsequent uprisings, as it created the sense that 'we are capable of managing these properties' (Wood 2003: 206).

Critiques of unequal access to land and livelihoods remained central to the guerrilla uprisings that led to the civil war in El Salvador; the majority of insurgents were subsistence farmers. Two years into the war, *campesinos* began taking land for their basic food needs, beginning with micro-plots (*microfundia*). Many stopped paying rent. Coffee plantations were pulled down for firewood and many estates were destroyed, especially those belonging to uncooperative landlords. While liberation theology did not cultivate specific discourses that critiqued the environmental devastation of Green Revolution technologies and the neo-colonial nature of development initiatives, it did provide a latent grammar for popular education movements that foregrounded these elements, especially the CaC movement, which re-emerged in El Salvador during the 1990s. The ways in which permaculture can be deemed a postcolonial AFN is discussed on p. 94 under the section entitled 'The knowledge politics of indigeneity'.

The first documented agroecology initiative in Central America was a small NGO programme in Guatemala in 1972, which aimed to empower a group of indigenous Kaqarikel *campesinos* by teaching ecological techniques that were embedded in their traditional culture (Holt-Giménez 2006). Projects centred on sharing long-standing principles for agriculture that were being eroded, such as recycling biomass, minimizing nutrient losses, and restoring degraded

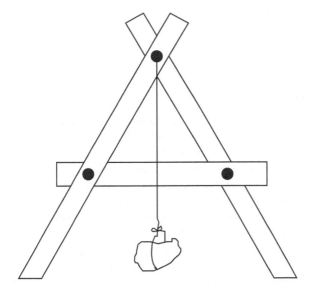

Figure 4.2 An *apparato 'A'* (A-frame).

soils. Bilingual *campesinos* were trained as farmer 'extensionists', and used simple instruments such as a machete, a tape measure and an *apparato 'A'* (a simple apparatus for measuring land gradients – see Figure 4.2), together with oral traditions, to communicate with other farmers. These were so successful that a number of encounters (*encuentros*) and exchanges (*intercambios*) were organized subsequently. Beginning in Guatemala and Mexico, these informal visits enabled farmers to consolidate and cross-fertilize their practices. The model gained momentum during the 1980s when heavy flooding exposed the difference between terraced farms, planted using traditional techniques, and modern farms, which were stripped of topsoil due to flooding. Whilst El Salvador was largely isolated from this growth, agroecological discourses entering the country as international non-governmental organizations (INGOs) were supporting village-level projects in conflict-affected areas during the early 1990s.

Crucial to agroecology and permaculture in this context is learning to experiment: each farmer or community of farmers is encouraged to test proposed techniques by constructing twin plots (*parcelas gemelas*) and measuring the differences in production. Each farmer is encouraged to become a promoter (*promotore*) for principles they find effective. As in liberation theology and post-colonial theory the emphasis is on rejecting received models of practice and legitimizing experience. Yet, in CaC, long-standing traditional practices and their infrastructures retain a significant role. It is maintained that viable futures cannot be devised through attention to genetic properties alone: these depend upon *cultural reservoirs* of knowledge and practices, shaped over many centuries. On the other hand, CaC pedagogies borrow from the lexicon of science to establish methodologies which emphasize repeatability and testing. From this perspective, agroecology marks a conjuncture between scientific principles for establishing translatable authority, and notions of tradition which resist notions of universal translatability.

Permaculture networks

The CaC movement reappeared in El Salvador via Nicaragua and Guatemala as leftist coalitions were breaking apart without the unifying mission of guerrilla warfare. Repeated efforts to repair fragmented economic infrastructures were foundering, whilst INGOs initiated aid interventions focusing on hunger relief and conflict resolution. Rising concerns in Europe with gender issues, discrimination against indigenous peoples and environmental degradation were reflected in these involvements and later in bilateral and multilateral funding (Pearce 1986). Meanwhile, a large number of new Salvadorian NGOs were founded in the 1990s, some with initial support from INGOs (such as Caritas, CARE International and Ayuda en Acción). Others received funding from external sources through their connection with liberation theology or bottom-up poverty alleviation. By 1989 there were around 700 NGOs in El Salvador, many of which provided opportunities for autonomous groups to operate outside state logics and jurisdictions (Pearce 1998). The influx of new streams of funding, however, created new kinds of divisions between classes and party factions, and the major INGOs became increasingly polarized by their entanglement with local political interests.

Emerging AFNs were thus mediated by political loyalties and access to international sources of funding. While a number of European and Australian donors promoted innovative regional developmental programmes that included local NGOs and community leaders, in practice older male brokers who had not necessarily been involved in guerrilla leadership ensured that benefits only reached those under their patronage (interview with Karen Inwood, director of the former Institute of Permaculture in El Salvador, 9 October 2012). Such dynamics are common in situations of high poverty and weak economic infrastructure, but served to further intensify the gendered and hierarchical structure of social organizing in El Salvador.

Agricultural practice formed a hotbed for such struggles. Besides a land redistribution programme mandated by the peace accords,[2] a rush of charitable investment from abroad framed itself through the emergent terminology of food security and the 'right to food'. Many of these investments did not meaningfully engage with community infrastructures, and tended to be short-lived. The vocabulary of agroecology, which had been developing in Central America since the 1970s, was consolidated to distinguish projects that emphasized bottom-up models of development, some of which also stemmed from international donors. Oxfam's South-to-South Programme of the 1990s, for example, drew on earlier CaC models to organize meetings between farmers across locations in Central America affected by conflict. Salvadorian farmers attended several of these meetings in Guatemala, and subsequently contributed to the development of an Eastern Commission and later a Western Commission of the Salvadorian CaC movement.

Salvadorian *campesinos*, already invested in agroecology, encountered permaculture practices through a series of brokers. Juan Rojas, a Salvadorian exiled during the civil war, had discovered permaculture in Australia, and returned to El Salvador in 1999 to become what he called a 'permaculture missionary'. Another important figure was Karen Inwood, a former community development worker from the UK, who discovered permaculture at a Scottish eco-village in 1999, which Rojas also attended. Inwood spent the following twelve years in El Salvador, using her community development background to implement permaculture as an extension of CaC activities. The Salvadorian Permaculture Association became established in 2001 as a project of FUNDHAMER (Fundacion Hermando Mercedes Ruiz), a para-church organization with access to funding from the Inter-American Foundation who supported bottom-up development projects. The group formally founded the separate Instituto de Permacultura de El Salvador (IPES) in 2002, with early grants from Misereor, a large German Catholic charity, Cordaid, a Dutch Catholic organization, the international Catholic aid organization CAFOD (Catholic Agency for Overseas development) and BMZ (Bundesministerium für Wirtschaftliche Zusammenarbeit und Entwicklung), a German federal ministry for poverty alleviation and development. Because of its early associations with liberation theology, IPES was able to attract funding from a range of organizations sympathetic to liberation theology as well as agroecology.

Permaculture in El Salvador is thus crucially tied to the renegotiation of spaces for autonomy and critique of top-down international interventions and developmentalism. However, permaculture is difficult to situate in terms of a postcolonial politics of knowledge. On the one hand, permaculture is tied to imported discourses and practices of agroecology and environmental design developed in the global North and perpetuated by international funding streams. On the other hand, permaculture resonates strongly with popular education practices that had already been developed in Central America, which work to disrupt knowledge hierarchies reproduced in state-led agriculture and international food security programmes. The main way permaculture can be seen to reinforce the latter dimension, and thus to unsettle its association with a dominant cultural politics, is by cultivating and legitimizing the expertise of the observer – in this case, the *campesino*. Rather than governmental technicians or project administrators, *campesinos* are given authority to design their own plots (*parcelas*) and to create their own networks to support these activities. Central to the design practices of a *campesino* are a set of twelve principles, each focused on the observation of how ecological and environmental processes take place. The idea is to create social systems that mimic the feedback and interactions of ecological networks comprised of diverse species communities and webs of activity. For example, the eleventh principle, 'use edges and values the marginal',[3] encourages designers to recognize how edges and boundaries may support the proliferation of pollinators, but also how in-between

Figure 4.3 Permaculture design course, Suchitoto, April 2014.

Source: Author's own photo.

spaces, such as hedges, can form valuable interfaces between agricultural *and* social factors. As illustrated by the guiding values of 'earth care', 'fair share' and 'people care', such practices are geared towards the creation of vibrant ecological and social relationships, including healthier soils.

  Through such practices, agriculture is reframed as a process of co-design or co-operation with existing communities, including more-than-human communities of microbes, animals and plants. Drawing strongly on the repertoire and ethos of popular education, permaculture achieves these aims through the use of simple instruments that can be mastered and taught by anyone. In each new location, a permaculture design course is run in partnership with existing local organizations to develop such teaching capacity (see Figure 4.3). This leads to the establishment of local permaculture associations, which form a basis for collective organizing. Subsequent stages in the strategic plan developed by IPES include establishing a nursery for the propagation of organic plants; the re-stimulation of local economies, especially through the reinvigoration of farmers' markets; and the establishment of long-standing structures for *campesino*-led collective learning and knowledge-sharing.

The knowledge politics of indigeneity

Permaculture principles are based around observation of, and participation with, already existing eco-social systems, and centre on design techniques which revive traditional and indigenous techniques. In this section the degree to which permaculture networks can be deemed an example of postcolonial AFNs is considered in two ways. First, the place of tradition in permaculture is explored, problematizing its new status within a transnational network of alliances while emphasizing the ways in which tradition works to protect diverse knowledge cultures from homogenization. This leads on to the idea of permaculture as a practical ethics for the production of postcolonial AFNs. Second, it is shown how this ethos leads towards the redefinition of principles for the governance of AFNs in a global South perspective, first by shifting emphasis from food security to food sovereignty, and then by reframing development as 'design'.

The place of tradition

In El Salvador, *'nuestros indígenas'* refers to the indigenous peoples who lived in the country prior to the Spanish invasion, and who perpetuated their own distinct cultural practices until the 1930s. A small indigenous minority of the Nahua-Pipil population still live in the south-western region of El Salvador, and there are smaller populations of Lenca and Cacaopera in the Eastern regions. However, most permaculturists are not themselves 'indigenous' in this sense. Rather, they draw on their mixed ancestry, identifying indigenous practices as those long practised in the region, and associated with indigenous ways of perceiving and interacting with plant and animal life. In reviving so-called 'lost' agricultural traditions that embody these forms of relationality, indigeneity becomes associated with connectedness to the land. For example, the stories, uses of plants and terminology remembered from indigenous communities are considered sacred knowledge to be preserved as an integral part of permaculture practices.

Yet this revalorization of indigenous knowledge draws on a mixture of research and reimagining, resulting in a hybrid set of practices. Rather than acquiring value solely by virtue of their longevity, appropriate permaculture practices are determined through notions of experimental testing and inter-cultural dialogue. Permaculture practices are traditional to the extent that they conserve the memory of indigenous practices into the present – yet they are open to modification through new methods and techniques. There is a certain pragmatism to tradition reimagined in this sense. That is to say, the principles of observation, sensing and listening that comprise the permaculture design repertoire are applied to evaluate long-standing agricultural practices, whether learnt from indigenous people from nearby or from another region. The criterion for deciding whether to adopt or retain a practice revolves around whether it works, but also whether it fits with the principles of inhabitation that derive from indigenous memory and practice. In this sense, permaculture can be understood as a practical ethics – it evaluates whether a practice *works* in terms of broader values of interconnected life systems. Such values reconstruct indigenous ontologies through *ecological* principles of species life, leading to agricultural practices that are good for food production but also good for the soil and for non-human animals.

Figure 4.4 Demonstrating the production of *bocashi*.

Source: Author's own photo.

An example of this practical ethos is manifest in the ongoing reinvention of the recipe for organic compost, known amongst Salvadorian permaculturists as *bocashi* (see Figures 4.4 and 4.5). *Bocashi* is an agroecological technique developed in Japan, which reached El Salvador via agroecological networks in Costa Rica and Nicaragua. To make *bocashi*, Efficient Microorganisms (EM) – mixed cultures of beneficial and naturally occurring microorganisms – are layered into compost matter with water and fermented over a month to increase the microbial diversity of soils and plants. The making of *bocashi* is an extremely important practice within the permaculture repertoire and is considered 'traditional' in relation to its connection with long-standing methods of creating fermented compost within the region. However it is also traditional in the sense that it benefits not only food production but the microbial life of the soil. Moreover, in reducing the labour needed to render soil fertile it diminishes the need for the application of compounds produced at other sites, in turn reducing energy input and financial costs. The exact method of heating compost in this way is relatively new to El Salvador, but the mode of restoring soil using ingredients that can be found close to hand (such as chicken and soldier ant waste, coffee grounds and molasses) is consistent with an indigenous ethos.

The hybridity of the resultant practices of permaculture is echoed at the global scale within growing food sovereignty movements. In particular, the notion of *Terra Madre* has been mobilized by diversely situated groups (including the Slow Food movement) to denote principles for ethical decision-making. In Salvadorian permaculture the term is mobilized to signal the way that permaculture design techniques are being reappropriated in alignment with indigenous histories and

Figure 4.5 Making *bocashi* compost.

Source: Author's own photo.

ontologies. The notion signals what I term a postcolonial ethos as it elicits a mode of organizing that is profoundly invested in the well-being of diverse ecological systems and sub-systems, and actively rejects forms of agricultural production grounded in a split between human-oriented systems and 'natural' systems. Most often the term is used to refer to the living vibrancy of the material world of plants, people and soil. It denotes the connectivity of life and, usually, the refusal of life to obey or even heed to human laws. The femininity of the principle is particularly important as the longer histories of global agriculture are seen to be closely tied to forms of appropriation – of both land and culture – that are typically white and male. Within emergent networks like permaculture and agroecology, *Terra Madre* is consequently becoming short-hand for a knowledge politics that is future-oriented, but premised in a reclaiming of both biodiversity as well as the ontological diversity on which that biodiversity is based. That is to say, ecological interconnectedness and flourishing is being associated with the ontologies, or 'worldings', of nature that colonial histories have violently suppressed.

In these terms, neither *Terra Madre* nor permaculture are able to be characterized in terms of one ontology. Indeed, as a permaculturist one can honour *Terra Madre* whilst also upholding a liberation theology *praxis*, or evangelical, Muslim or atheist beliefs. Rather than carrying their own ontological content permacultural practices are comprised of ethical and pragmatic principles that allow collectives to (re)design environments in dialogue with the more-than-human world, via the ontological forms of knowing and traditions that are practised there. This ontological empty space is what makes the principles of permaculture so valuable for learning in different contexts, as it provides an opportunity for knowledge production that counteracts imposed universals.

The hybridity of this process is also vital. In practice, as my interviewees told me, it is normally only well-meaning NGOs who insist that indigenous practice is something that is or ever was something 'pure' and separate from all other forms of knowledge – a claim which some suggest in fact relegates indigenous peoples to the status of past cultures (Braun 2002), along with the 'pristine' forest (Denevan 1992). Indeed, whilst it is important to recognize that, historically, Western onto-logies of nature and culture have violently colonized other ways of knowing the more-than-human world, in the Salvadorian context the idea of a pure underlying substrate of 'indigenous knowledge' misrepresents the constellation of cultural ecologies which informs contemporary *campesino* cultures. It also eclipses the way that *Terra Madre* is increasingly mobilized within food sovereignty framings as a connective concept allowing for cultural exchange between diversely posi-tioned land-workers and agroecological practitioners. Like food sovereignty (see Martinez-Torres and Rosset 2010), *Terra Madre* carries something of a strategic essentialism without itself being an essentialist concept. It forms a vehicle with which to translate diverse experiences and cultural commitments, without requir-ing that they compromise their autochthonous singularities. Through emergent networks like permaculture, tradition is becoming reinvented as an ethical framing for agriculture through which one may protect diverse ontological accounts of 'nature' as well as diverse forms of biological life.

Food sovereignty and design

This reinvention of tradition clearly has important impacts on the way that food security and development are imagined within alternative food systems, as it ties together notions of the bottom-up governance of agricultural production with the protection of ancestral practices and the deconstruction of pervasive knowl-edge hierarchies. Within El Salvador these terms are increasingly being replaced by the language of food sovereignty and design. These operating principles are two key distinguishing features of postcolonial AFNs in this context, as they embed the politics of knowledge associated with a traditional ethos into future-oriented forms of production and consumption. This takes place by a resituation of food *producers* as the experts of food production, and a rejection of knowledge regimes which legitimize other knowledge-power hierarchies. Tradition claims an intimacy with more-than-human agencies and a consequent authority to speak that reframes *campesino* farmers as experts and protagonists of food futures. Meanwhile, the emphasis in agroecology on *in situ* expertise establishes these claims through the repeatability and translatability of scientific discourse. What unifies such movements in this sense is not one ontology, as in cases where human rights are expounded as a universal solution, but a commitment to diversity: to biodiversity, but before this, to ontological diversity, without which biodiversity cannot be achieved.

The shift from food security to food sovereignty reflects the globalizing discourses discussed in earlier sections, but, for postcolonial AFNs it is also important that this terminology connects 'alternatives' to land politics. Lucas Argueta, agricultural technician for the Salvadorian NGO Fundacíon Segundo Montes,[4] made this connection clear when he reflected on the emergence of permaculture in El Salvador:

> Food security is about access and quality of food, but it is also about access to *land* . . . It is about land ownership and land governance, and through this, protecting the understanding we have of the diversity of products in the land. And health is not just about nutrition, although that's also important. It is fundamentally tied to the contamination of the soil and capacities for regenerating it. Soil quality is this [*picks up a banana from his desk*]: food. And so food security is primarily a relationship with the land . . . and we cannot protect the land without the autonomy of the *campesinos* themselves.
>
> (Interview with coordinators of Segundo Montes, 20 April 2014)

Here Argueta presents food security *as* a form of food sovereignty, implicitly rejecting other definitions determined 'from above'. For Argueta, as for others interviewed, food security is impossible without a legitimization of the *campesino* experience and expertise in relation to the land. Food security is impossible without food sovereignty, and cannot be separated from dialogue over land ownership or the country's revolutionary history. Consequently the construction of AFNs that embody a postcolonial ethos is tied to a form of food production that also legitimizes bottom-up knowledge production and critique.

Thinking about food sovereignty in this way means addressing the rising authority of NGOs in the Salvadorian context, and the influx of alternative models (including permaculture) that may also presume to carry a solution. These models may have much to offer, but it is critical to explore how to open dialogue over possible structures without 'speaking for' others, as Mauricio Fuente, director of the agricultural NGO Plateforme suggested:

> Although agriculture is a political theme for us here, we as organizations can't keep regarding ourselves as the protagonists of the struggle. We have to understand that it's the *campesinos* who are at the forefront of this kind of struggle, who are developing this kind of agriculture, out of their own experience. So our strategies are to try and connect these forms of struggle, not tell people what to do.
>
> (Interview with Mauricio Fuente, 12 April 2014)

Sixto Hernández, chief executive of Segundo Montes, insists that for these reasons education – specifically popular education principles – must be the medium for the construction of food alternatives. For him, food sovereignty is fundamentally premised in a valorization of skills and understandings that have long held agriculture together, which are undermined by the large-scale, export-oriented models favoured by the World Bank and International Monetary Fund. From Fuentes' point of view it is critical that an example like permaculture is not treated as the only means of constructing alternative economies. Fuentes drew attention to a plethora of other networks within El Salvador that are adapting CaC or other agroecological pedagogies to form their own networks and systems of production. What is crucial for him is not the approach *per se* (permaculture or another) but the common way all of them have of problematizing the situation: the need to produce new food publics that foster disagreement and debate about food governance (my terms).

The focus on democratic deliberation reflects a broader shift in terminology from development (*desarollo*) interventions to processes of design (*diseño*). In permaculture, the term 'design' is used to describe the process of creating a food production system or even a social or distributive system through observation and interaction with other existing systems, such as climatic micro-systems (see Figure 4.6). It highlights growers, or in this case *campesinos*, as those who come up with design solutions, as experts in the particularities of land and existing cultures of production. In the context of emergent regional debates, and in some cases national processes of decision-making in El Salvador, the term is also being mobilized to discuss platforms for exchanging qualitatively different forms of expertise. Thus permaculture practitioners have been invited to participate in creating rural development strategies on the basis that 'they are keeping the know-how [*experiencia y conocimientos*] of our ancestors alive, *and* the bodies of our

Figure 4.6 Modelling micro-climate systems.

Source: Author's own photo.

people more healthy' (interview with Michael Gusman, 19 April 2014). Whilst permaculture communities have different interests in health, this common area of concern has provided a dialogical sphere that enables the translation of different forms of expertise. As such, the language of health has provided an important vehicle for articulating a permaculture ethos elsewhere, without diminishing the emphasis on bottom-up knowledge creation.

The shift towards design challenges the ways that development-oriented food solutions tend to site expertise in the know-how of governmental technicians or NGO administrators, who mediate projects from above. In line with the principles of design, agricultural technicians in El Salvador have been partaking in perma-culture training. Flor de Flores, an agricultural technician at the Fundacíon Dan Vicente Productivo in San Vicente, explained how permaculture was in fact being used to change the practices of technicians. De Flores, one of three technicians studying for a Permaculture Design Certificate in Suchitoto in 2014, suggested that the role of technician is being reframed by permaculture principles, giving them the opportunity to reimagine their work as well. Using simple instruments, a technician can encourage a *campesino* farmer to determine their own solutions, thus reducing dependency on governmental strategies.

This legitimization of *campesino* expertise, which I am associating with a postcolonial politics of knowledge, also influences how people design their living environments, including agricultural plots (*huertos*), domestic spaces, communal spaces and life patterns. By changing the way that permaculture practitioners view their own intelligence and capacities, permaculture fosters design innovation that reframes space and efficiency in terms of connectivity between systems and long-term health (thought of in terms of the health of plural systems, not just human health), rather than short-term profit. Thus Armando López, an environment officer in Suchitoto, positioned *bocashi* as one of the most *political* products of permaculture. From López's perspective, learning this technology produces autonomy for families, by saving them money, but also by empowering them to diagnose and remedy problems with their own soil and yield. While the practice mirrors the hybridity discussed in the previous section, it also embodies a reclamation of food sovereignty through the revitalization of bottom-up networks of knowledge and production. Through this process, explained López, social critiques that were once targeted at coffee élites within El Salvador take on new, *global* targets: large multi-nationals such as Monsanto, institutions such as the World Bank. Revitalizing agricultural traditions in dialogue with agroecological practices from other places has allowed for postcolonial critiques of power that extend beyond borders.

Conclusions

In this short chapter I have only brushed the surface of the complexity of a post-colonial politics of knowledge in relation to AFNs in El Salvador. In tracing the emergence of the permaculture network in El Salvador I have, however, drawn out three important findings that I suggest can contribute towards an emergent understanding of AFNs in this as well as other contexts. First, I have emphasized the critical importance to the development of AFNs of prior networks of popular education and collective organizing that keep a postcolonial politics of knowledge central to their modes of organizing. Whilst drawing strongly on models and funding streams from the global North, the permaculture network in El Salvador has insisted upon, and further elaborated, the centrality of indigenous and traditional knowledges to their practice, and in this way has shaped a permaculture practice that is different from its 'Western' equivalents. It is extremely important to recognize this heritage so as not to dismiss permaculture as itself a cultural imposition or to valorize design techniques without acknowledgement of the importance of long-standing cultural contexts.

This links to the second important contribution of this chapter: that what enables the perpetuation of a postcolonial knowledge politics through national networks, and indeed transnationally, is a reinvention of tradition itself as a form of ontological empty space. In common with networks such as LVC, permaculture mobilizes figures like *Terra Madre* to connect diverse indigenous approaches to nature, without insisting on their purity. This allows for hybrid forums of exchange, grounded in a pragmatic and experimental approach, which hold in common their rejection of universalizing claims to knowledge about environments. As such they embody an alternative ethos that insists upon the creation of solutions in dialogue with (a) historical experiences of oppression, (b) first-hand experience of the land and its processes and (c) close observation of diverse ecological systems. Permaculture, as a practical ethics, insists that biodiversity and ontological diversity cannot be separated.

Finally, and to conclude, I have emphasized that what characterizes postcolonial AFNs in El Salvador – and what characterizes them as such – is the grounding of food alternatives in a legitimization of *in situ* expertise that shifts the terms of food poverty debates away from food security and development and towards food sovereignty and design. What is postcolonial about such interactions is the rejection of the basic premise of global food production systems premised upon the division of those who think and design from those who labour and enact. What is alternative about the new food networks that are emerging is perhaps not that they are entirely 'local' or somehow purified from capitalist systems of exchange, rather that the solutions they promote are grounded in an invitation to intimacy with more-than-human ecologies and relationships. In order to take part in such an alternative, both humans and non-humans must become participants.

Notes

1 Plant-hunters are those who collect live plant specimens from the wild. A practice that has occurred for centuries, plant-hunting in this sense refers to scientists and their allies who sought out exotic plants to bring back to colonial centres like England, especially during the eighteenth and nineteenth centuries. Such specimens would often end up in botanical gardens or the private gardens of wealthy collectors.

2   El Salvador's 1992–7 Land Transfer Program, known as El Programa de Transferencia de Tierras, was mandated within the Chapultepec Peace Accords concluded on 16 January 1992. The Program laid out principles for legalizing tenancy in occupied conflict zones, although its ambiguity led to considerable tensions in the following decade.
3   See www.permaculture.org.uk (accessed on 15 April 2016).
4   The foundation, established in 1993, works with 150 families in North Morazán on themes including agricultural production and women's organizations, creating a solidarity economy and institutional sustainability.

References

Agrawal, A. (1995) 'Dismantling the divide between indigenous and scientific knowledge'. *Development and Change*, 26(3): 413–39.
Allen, P., M. FitzSimmons, M. Goodman and K. Warner (2003) 'Shifting plates in the agrifood landscape: the tectonics of alternative agrifood initiatives in California'. *Journal of Rural Studies*, 19(1): 61–75.
Braun, B. (2002) *The Intemperate Rainforest: Nature, Culture, and Power on Canada's West Coast*. Minneapolis: University of Minnesota Press.
Cabarrús, C. R. (1983) *Génesis de una revolución: Análisis del surgimiento y desarrollo de la organización campesina en El Salvador*. Mexico City, DF: Centro de Investigaciones y Estudios Superiores en Antropología Social.
Carey, D. (2009) 'Guatemala's green revolution: synthetic fertilizer, public health, and economic autonomy in the Mayan highland'. *Agricultural History*, 83(3): 283–322.
Chakrabarty, D. (2009) *Provincializing Europe: Postcolonial Thought and Historical Difference*. Princeton, NJ: Princeton University Press.
Denevan, W. M. (1992) 'The pristine myth: the landscape of the Americas in 1492'. *Annals of the Association of American Geographers*, 82(3): 369–85.
Descola, P. (2010) 'Steps to an ontology of social forms'. In T. Otto and N. Bubandt (eds) *Experiments in Holism: Theory and Practice in Contemporary Anthropology*. Malden, MA and Oxford: Wiley-Blackwell, 209–26.
DuPuis, E. and D. Goodman (2005) 'Should we go "home" to eat? Toward a reflexive politics of localism'. *Journal of Rural Studies*, 21(3): 359–71.
Edelman, M. (1998) 'Transnational peasant politics in Central America'. *Latin American Research Review*, 33(3): 49–86.
——(2005) 'Bringing the moral economy back in … to the study of 21st-century transnational peasant movements'. *American Anthropologist*, 107(3): 331–45.
Fairbairn, M. (2014) ' "Like gold with yield": evolving intersections between farmland and finance'. *The Journal of Peasant Studies*, 41(5): 1–19.
Freidberg, S. and Goldstein, L. (2011) 'Alternative food in the global south: reflections on a direct marketing initiative in Kenya'. *Journal of Rural Studies*, 27(1): 24–34.
Goodman, D. (1999) 'Agro-food studies in the "age of ecology": nature, corporeality, bio-politics'. *Sociologia Ruralis*, 39(1): 17–38.
Goodman, M. K. (2004) 'Reading fair trade: political ecological imaginary and the moral economy of fair trade foods'. *Political Geography*, 23(7): 891–915.
Guha, R. (1999) *Elementary Aspects of Peasant Insurgency in Colonial India*. Durham, NC and London: Duke University Press.
Harris, E. M. (2010) 'Eat local? Constructions of place in alternative food politics'. *Geography Compass*, 4(4): 355–69.

Heynen, N. (2010) 'Cooking up non-violent civil-disobedient direct action for the hungry: "Food Not Bombs" and the resurgence of radical democracy in the US'. *Urban Studies*, 47(6): 1225–40.

Hinrichs, C. (2003) 'The practice and politics of food system localization'. *Journal of Rural Studies*, 19(1): 33–45.

——(2007) 'Introduction: practice and place in remaking the food system'. In C. Hinrichs and T. Lyson (eds) *Remaking the North American Food System: Strategies for Sustainability*. Lincoln: University of Nebraska Press, 1–15.

Holt-Giménez, E. (2006) *Campesino a Campesino: Voices from Latin America's Farmer to Farmer Movement for Sustainable Agriculture*. Oakland, CA: Oakland University Press.

Hume, M. (2009) *The Politics of Violence: Gender, Conflict and Community in El Salvador*. London: Wiley-Blackwell.

Jarosz, L. (2008) 'The city in the country: growing alternative food networks in metropolitan areas'. *Journal of Rural Studies*, 24(3): 231–44.

Leyshon, A., R. Lee and C. Williams (eds) (2003) *Alternative Economic Spaces*. London: Sage.

McMichael, P. (2014) 'Historicizing food sovereignty'. *The Journal of Peasant Studies*, 41(6): 933–57.

Makki, F. (2014) 'Development by dispossession: Terra Nullius and the social-ecology of new enclosures in Ethiopia'. *Rural Sociology*, 79(1): 79–103.

Mangelsdorf, P. C. (1951) 'Hybrid corn: its genetic basis and its significance in human affairs'. In L. Dunn (ed.) *Genetics in the 20th Century: Essays on the Progress of Genetics during its First 50 Years*. New York: Macmillan Press, 555–72.

Martinez-Torres, M. and P. Rosset (2010) 'La Vía Campesina: the birth and evolution of a transnational social movement'. *The Journal of Peasant Studies*, 37(1): 149–75.

Montes, S. and M. Gaibrois (1979) *El Compadrazgo, Una Estructura de Poder en El Salvador*. San Salvador, El Salvador: UCA editores.

Montobbio, M. (1999) *La Metamorfosis de Pulgarcito: Transición Política y Proceso de Paz en El Salvador*. London: IIED Press.

Pearce, J. (1986) *Promised Land: Peasant Rebellion in Chalatenango, El Salvador*. London: Latin America Bureau.

——(1998) 'From civil war to "civil society": has the end of the Cold War brought peace to Central America?'. *International Affairs*, 74(3): 587–615.

Pimbert, M. (2006) *Transforming Knowledge and Ways of Knowing for Food Sovereignty*. London: IIED Press.

Purdue, D., J. Dürrschmidt, P. Jowers and R. O'Doherty (1997) 'DIY culture and extended milieux: LETS, veggie boxes and festivals'. *The Sociological Review*, 45(4): 645–67.

Roseberry, W. (1991) 'La falta de brazos'. *Theory and Society*, 20(3): 351–82.

Smith, C. (1991) *The Emergence of Liberation Theology: Radical Religion and Social Movement Theory*. Chicago: University of Chicago Press.

Spivak, G. (1985) 'Can the subaltern speak? Speculations on widow-sacrifice'. *Wedge*, 7/8: 120–30.

Viveiros de Castro, E. (2004) 'Perspectival anthropology and the method of controlled equivocation'. *Tipití: Journal of the Society for the Anthropology of Lowland South America*, 2(1) [online]. Available at http://digitalcommons.trinity.edu/tipiti/vol2/iss1/1 (accessed on 6 May 2015).

Whatmore, S., P. Stassart and H. Renting (2003) 'What's alternative about alternative food networks?'. *Environment and Planning A*, 35: 389–91.

Wittman, H. (2009) 'Reworking the metabolic rift: La Vía Campesina, agrarian citizenship, and food sovereignty'. *The Journal of Peasant Studies*, 36(4): 805–26.

Wood, E. J. (2003) *Insurgent Collective Action and Civil War in El Salvador*. Cambridge: Cambridge University Press.

5 Possibilities for alternative peasant trajectories through gendered food practices in the Office du Niger

Nicolette Larder

Introduction

This chapter explores the existence and nature of mainstream and alternative food networks (AFNs) in the Office du Niger, Mali in the context of women's historical marginalization from land and farming. The Office du Niger is the name given to a designated rice-growing zone in Mali that was first developed by French colonialists in the 1920s, who envisaged the region as a massive cotton growing operation that would supply European markets (see Figure 5.1). Following Malian independence in 1960, the Office came under the control of the Malian government, who set up an autonomous body, known locally as 'the Office', in order to manage the region. After financial deregulation during the 1980s and 1990s, the role of government in the region was much reduced although the state retained ultimate control over the region and its lands. Today the Office administers approximately 96,000 hectares of irrigated rice land farmed primarily by smallholder farmers. Cultural denigration of non-rice production in the zone, set in place almost a century ago, continues to inflect thinking around the legitimacy of food practices. The colonial legacy has been the construction of irrigated rice as the mainstream, legitimate crop of the Office du Niger. Production on non-irrigated land, whether for subsistence or sale, is not recognized as lawful or purposeful.

This chapter focuses on the ways in which historical gender relations have shaped participation in, and definitions of, contemporary mainstream and alternative food networks in the Office. From its early development, the region has been associated with men: it was conceived by male European technocrats and built by 50,000 men conscripted from across French West Africa (Twagira 2014). While early colonial policy encouraged families to settle in the region, prior to 1940 very few women and children lived there due to the area's harsh living and labour conditions (Twagira 2014). While women and children have been settled in the region since the 1940s, according to local women the Office is still 'for men', a social fact related to men's ongoing privileged access to irrigated rice lands and rice-related technologies. Men's tools include threshing machines, tractors, scales, synthetic fertilizers and hybrid seeds, while women's tools include hand-held ploughs and organic fertilizer made from kitchen waste. The masculine nature of mainstream rice production has shaped the nature of alternative food practices in

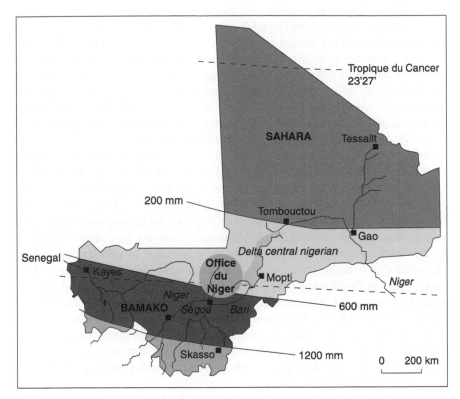

Figure 5.1 Location of the Office du Niger, Mali.

Source: Coulibaly *et al.* (2006).

the region, which exist as 'alternative' precisely because of their relationship to mainstream rice production. In line with Abrahams' (2007) work on AFNs in Johannesburg, this chapter suggests that participation in AFNs in the Office is about enhancing livelihoods in the face of women's marginalization from the mainstream, rather than a political or lifestyle choice. As a result, AFNs in the Office are not necessarily restricted to a range of practices in opposition to the mainstream, as the dominant Northern conception of AFNs would suggest (Abrahams 2007). Rather, AFNs in the Office may incorporate or seek to move towards, rather than away from, mainstream practices, as poor and marginalized individuals and families seek to enhance their livelihood strategies.

The findings of this chapter have implications for recent moves by the Malian government and peasants' rights groups in Mali towards food sovereignty. The rise of the transnational peasant movement, *La Via Campesina*, alongside the growth of alternative food movements in the minority world has challenged the existing productivist model of the global food system. Today, the model of food sovereignty, which prioritizes the rights of people to define their own agriculture and food policies (Via Campesina 2001, cited in Jarosz 2014), is recognized by many as a viable and worthwhile alternative to the extractive industrial food system (McMichael 2006). Diversity and self-determination at the grassroots level are central tenets of the food sovereignty model. However, because grassroots practices are situated in place and grounded in particular historical and socio-cultural contexts, I suggest that they may diverge from an ideal 'alternative' which sets itself against a 'mainstream' (see Chapters 1 and 2, this volume). As Abrahams (2007) argues, AFNs in the global South share commonalities with those in the North – primarily in counteracting conventional food networks – yet AFNs in the South have distinctly different drivers. The findings of this chapter suggest that these drivers matter, in all their complexity, when working towards food sovereignty. In particular, as food sovereignty travels it takes on different conceptualizations and understandings that do not always bear resemblance to the idealized version. The slipperiness of the terms 'alternative', 'mainstream' and 'sovereign' means that they can be re-cast again and again across place and time (Boyer 2010). Moreover, grassroots participation in AFNs in the South cannot always be read as an expression of political opposition mainstream or a desire for an expansion of resistant ideologies. Rather, AFNs may be part of livelihood strategies that reflect moves towards the mainstream. Indeed, as I argue for the case study of the Office du Niger, in some cases AFNs exist due to an inability to access conventional food systems, rather than an ability to exercise choice. As a consequence, this chapter raises questions about how historical and grounded experiences of agriculture and farming might shape the nature of the emergent food sovereignty movement in Mali and the global food sovereignty movement more broadly.

Data included in this chapter were collected in 2011 in the Office du Niger. The research involved two trips to the region: the first was carried out in February 2011 over a ten-day period and the second was in August 2011 for a two-month period. Over the course of my time in the region, I undertook interviews and focus groups with around 200 individuals, mostly smallholders but also locally-based Office du Niger staff. Living in the region provided an opportunity to observe village life and participate in daily food practices including production, processing and consumption and to observe a range of relevant aspects of life including household dynamics, market days and important rituals including the end of Ramadan and the National Day of Independence.

First, the chapter explores the historical development of agriculture in the Office du Niger during colonial rule, situating the roots of contemporary mainstream and AFNs in this period. Second, recent moves in Mali towards food sovereignty are discussed, examining how colonial legacies have influenced the resultant version of food sovereignty in the country. Finally, practices associated with mainstream and AFNs in the Office du Niger are explored, including women's desires for incorporation into the mainstream and the implications for the emergent food sovereignty movement in Mali.

Colonialism, the Office du Niger and the emergence of mainstream food practices

Agriculture in the Office du Niger is rooted in the French colonial period as the French pursued extra-territorial production of food and other staples for the metropolis. Although France had held the posts of Gorée and Saint-Louis in Senegal since the 1600s (Aldrich 1996), most French colonization in Africa occurred during the 1800s, in North Africa in the mid-1800s and in West Africa in the late 1800s. One result of this colonization was the doubling of trade between West Africa and France between 1885 and 1901 (Suret-Canale 1971). Agricultural-based developments in the Niger Delta in the region of the Office du Niger exploited the colony's natural resources, with trials conducted to test the suitability of different crops for West African conditions. Similar to French architects who saw urban West Africa as a *tabula rasa* for new urban designs, in West Africa land was seen as a laboratory for agricultural experimentation (Osborn 2011). Nineteenth-century correspondence between Dakar (the seat of French Sudan power) and Paris indicates that various crops were trialled for their suitability for export. In the first decade of the twentieth century, products such as oil, peanuts, cotton, tobacco, rice, wheat, potatoes, beans and numerous varieties of seeds and building materials were trialled in the region.

As part of this 'perfecting' of agriculture in the region, the French sought to reform land rights and social customs. One of the most significant changes to land laws in West Africa was the implementation of the 1935 Forestry Decree, which gave the colonial government power over natural resources by classifying all land as 'empty' or without an owner (Benjaminsen 1997: 130). European concepts of property dictated how land 'should be' used. For example, any land that was in long-term fallow or used for foraging was taken over by the state. However, the French had difficulty imposing European ideas of ownership on the people of French Sudan; until that time land use had been predicated on usufruct rights where village land was communally owned with use rights granted to families by the village chief (Becker 2001). Concepts of land rights imposed by the European colonizers conflicted with local understandings about land and nature. These conflicts were evidenced in colonial and postcolonial relations between peasants and the state, as '[p]easant land-use practices and rules came into conflict with the written texts of colonial law, setting in motion social and political contradictions that remain unresolved' (Becker 2001: 510). The French colonizers also believed that changes to social norms and customs were essential for creating a class of obedient and disciplined workers, especially given the central role of local labour for the exploitation of natural resources. While some have suggested that this manipulation was based on emancipatory ideals – Africa was a primitive and homeostatic place that 'needed' development (Suret-Canale 1971) – any emancipation was arguably designed to create a happier (and hence more controllable) colonial subject (Osborn 2011).

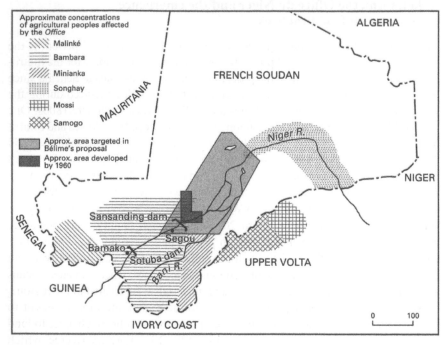

Figure 5.2 Location and extent of the Office du Niger under French rule.

Source: Filipovich (2001).

The development of the Office du Niger

It was within this context that French colonial powers developed the Office du Niger (see Figure 5.2). The Office was the brainchild of Emile Bélime, a US-born hydrologic engineer who, familiar with other irrigation schemes in the French colonies, convinced the French authorities to begin an irrigation scheme in the Niger Delta (Morabito 1977). In 1912, Bélime presented a plan for a large-scale irrigation project, in the order of 1,850,000 hectares in the Niger Delta of French Sudan, to the Governor-General of French West Africa (Filipovich 2001). The Office was originally designated as a cotton-growing region, producing both Egyptian and American cotton varieties, and was part of a plan to increase cotton production in the West African Sahelian region (Church 1951). At the turn of the nineteenth century, Europe relied almost entirely on America for its supply of raw cotton, and expansion of American processing facilities threatened European supply, particularly during the First World War (Filipovich 2001; Benjaminsen *et al.* 2010). Cotton production in the Office was therefore aimed at reducing France's dependence on American cotton.

'Robot farmers' under colonial rule

One of the primary tasks for the architects of the Office was the creation of a labour force. The French vision for the Office was a dehumanized labour force, which the French called 'robot farmers'. While oral histories suggest families were living and farming the region prior to development of the Office, the French plan for an expected 960,000 hectares of irrigated land required around 300,000 workers. The French, under the persuasion and direction of Bélime, attempted to fill the 'underpopulated' lands with labour through coercion and by force, with men conscripted from neighbouring areas, including present-day Niger, North Africa and Upper Volta (present-day Burkina Faso) (Morabito 1977; Filipovich 2001). The first settlers were installed in newly created villages in the Office in the mid-1930s, each populated with around 300 settlers (Filipovich 2001). Settlers included the Bambara people from French Sudan and the Mossi people from Upper Volta (Songre 1973; see Figure 5.2). Men worked the land with new farming methods – ploughs rather than hoes and intensified farming techniques – and lived with their families on privately owned irrigated plots (Twagira 2014).

Numbers of voluntary settlers in the region grew slowly, despite propaganda campaigns by the Office du Niger that promised tax reductions or exemptions for new settlers (Filipovich 2001). Unsurprisingly, a review of the operations of the Office published in 1944 suggests that the number of families who voluntarily moved to the region was well below the numbers who were forcibly moved there. A lack of labour saw the Vichy government of France revise the initial size of the irrigated area from 960,000 hectares to 160,000 hectares and by 1934, only 3,000 hectares of land was under production, producing groundnut, rice, millet and cotton for export. By 1941 this area had increased by only 15,000 hectares – less than 2 per cent of the original area designated for export production (and less than 10 per cent of the revised plan) (Morabito 1977). By the end of the Second World War, the French faced significant difficulties increasing production in the Office to reach desired volumes (Morabito 1977).

Growing diversity under colonial rule

Beyond the refusal of people to migrate to the region, there was significant reluctance and resistance on the part of those who were forcibly settled in the region to grow the colonial crops of cotton and rice. It is in this resistance that we see the emergence of what I call AFNs. Local oral histories confirm that the French administration exerted considerable power over the labour force by limiting settlers' ability to leave the region, enforcing the rule that all inhabitants had to be in the fields between sunrise and sunset and enforcing the number of work hours per day (Filipovich 2001). With peasants forced to produce export crops rather than subsistence crops, hunger and food insecurity became widespread throughout the region; French-supplied rations were insufficient to meet local food needs and a preference for millet over rice exacerbated resistance to mainstream agricultural practices (Twagira 2014). Aggravating the problem was the fact that there were so few women in the region. Women in Mali were, and continue to be, responsible for cultivating fruit and vegetables, and meal preparation. As explained below, Malian women are responsible for providing the taste element of grain-based meals. By the mid-1930s the French administration realized that women were required if the region's population was to grow and began to bring in hundreds of families in an attempt to increase the number of women and children living and working in the Office du Niger (Twagira 2013).

While men had cultivated subsistence crops alongside cotton and rice, such agricultural practices increased with the entrance of women to the area (van Beusekom 2002). In the mid-twentieth century, peasants placed more value on subsistence crops than export crops, working on rain-fed subsistence crops first with the use of state-allocated fertilizers, which were designated for irrigated crops (van Beusekom 2000). This value of subsistence over export crops was not unique to the Office du Niger and also occurred in cotton-growing regions in southern Mali (Benjaminsen *et al.* 2010). The French administration sought to discipline workers who privileged subsistence crops in this way and reportedly attempted to evict settlers who failed to attend to their irrigated fields (Twagira 2013). By the mid-1950s the French goal of 'civilizing' West Africa was slipping out of reach and French politicians and the media began to question the cost and benefits of 'development'. As a result, France's original argument for colonization – Africa was poor and backward – was turned into the reason for decolonization (Cooper and Packard 1997).

In 1960 Mali was granted independence from France and the newly created Republic of Mali took over administration of the Office du Niger with Isse Ongoïba given directorship of the zone. Following French leaders, Ongoïba sought to forcibly end non-rice production: banning farmers from working on crops other than irrigated rice, instituting a tax on subsistence crops and intensifying surveillance by local police (Twagira 2013). Today, non-irrigated production retains its air of illegitimacy in the zone, with irrigated production the only kind of practice sanctioned by the state. With this historical context in mind, the next section explores recent moves in Mali towards a food sovereignty model and shows how colonial legacies have influenced this vision.

Locating food sovereignty in the Office du Niger

Food sovereignty and the Malian State

As Millner (Chapter 4) illustrates, food sovereignty has come to be a driving agenda of numerous peasant rights organizations, many who coalesce under the banner of *La Via Campesina*. The right to land for those who work it, and unity in diversity, are central in the food sovereignty model (Wittman 2011). Within *La Via Campesina*, diversity is conceptualized in relation to gender, class and ideology. For example, since its inception there have been strong mandates from within *La Via Campesina* for gender equality; the movement has argued that food sovereignty can only be achieved by ending violence – understood in its broadest sense – against women (Andrée *et al.* 2014). Diversity is also conceptualized in terms of class relations between 'small' and 'large' farmers, and in different understandings of the ways in which peasants address social and environmental challenges posed by the contemporary food system (Wittman 2011). While predominately a producers' movement, many activists and academics in the minority world now advocate for food sovereignty. For these actors, the rights-based food sovereignty concept overrides exclusionary and consumption-focused AFNs of the North (Desmarais 2004; Fairbairn 2012). To this end, food activists in the North are incorporating food sovereignty into their lexicon and reframing food-based activism away from alternative consumption and towards increasing rights for food consumers *and* producers (see, for example, Food Secure Canada 2011; Australian Food Sovereignty Alliance 2013). At the same time, *La Via Campesina* is actively growing its network of national organizations affiliated with the movement.

Mali is one of the few states in the world to recognize food sovereignty at the national level. In 2005 Mali's national coordination of peasant organizations (CNOP) became the first organization in West Africa to join *La Via Campesina*. Since that time, CNOP has continued to advocate for farmers' rights in Mali with a particular focus on the achievement of food sovereignty. At the same time that CNOP became associated with *La Via Campesina*, the Malian Government officially incorporated food sovereignty into law. In 2006 the National Agriculture Law (Loi d'Orientation Agricole; LOI) was passed in Mali, which claims to 'guarantee food sovereignty' for Mali (Republic of Mali 2006).

On the surface, this apparent congruence between state and peasant propositions for food sovereignty suggests the possibility of reshaping food and agriculture systems in Mali along more socially and ecologically appropriate lines. However, the Malian government's conceptualization of food sovereignty differs markedly from that of the idealized *La Via Campesina* version. Food sovereignty, as defined in the LOI is: 'the availability and accessibility at all times of quality food products for the satisfaction of domestic food needs, based primarily on national agricultural production following nutritional and local culinary practices'(Republic of Mali 2006). This definition reads as a combination of food security and food self-sufficiency, rather than the rights-based version of food sovereignty upheld by *La Via Campesina*. At the same time, the LOI draws on market liberal and productivist discourses to define the country's agricultural development strategy, stating that 'the development strategy of agricultural production is focused primarily on the spatial measures of intensification [and] diversification and sustainability of production according to comparative advantages [and] competitive products' (Republic of Mali 2006).

Moreover, the Malian State has been the target of some of the most significant social movement actions against the so-called African 'land grab' of recent years. In late 2008 civil society organizations in Mali began to report on the marked increase of foreign investments in the Malian agricultural sector, particularly in the Office du Niger. While information on the amount of land under contract to foreign entities is not systematically recorded by government agencies in Mali, on-ground research by Diallo and Mushinzimana (2009) suggests that the Malian government approved seven large-scale projects by foreign investors between early 2004 and late 2009, totalling 162,850 hectares. The reports suggest that the vast majority of these investments were made in the Office du Niger.

The largest of these investments was the Malibya project, which involved a ninety-nine-year lease of 100,000 hectares of land in the Office to Malibya, a subsidiary of the Libyan Sovereign Wealth Fund. The first stage of the project involved the development of a forty-kilometre canal and road in the Office, which destroyed gardens and houses in several villages (Oakland Institute 2011). While land leases to foreign entities authorized by the Malian government are con- demned by CNOP, the government justifies its actions by arguing that foreign investment is required to expand irrigation networks in order to achieve 'food sovereignty' in line with the LOI (Larder 2015). By contrast, *La Via Campesina* use the highly political term of 'land grabbing' to problematize such actions, a term that has been adopted by scholars and activists in recent years (for example, Hall 2011; Borras and Franco 2012). Central to *La Via Campesina*'s argument against land grabbing is that the practice fundamentally undermines Mali's ability to achieve food sovereignty because it denies farmers' rights to land (La Via Campesina 2012). Since 2009 *La Via Campesina* and CNOP have been engaged in a high profile anti-land grabbing campaign on behalf of farmers affected by land grabbing in the Office du Niger.

All of this suggests that incorporation of the once-fringe discourse of food sovereignty into national agricultural law in Mali has not occurred seamlessly, and that the Malian government's version of food sovereignty diverges signifi- cantly from *La Via Campesina*'s vision. Arguably, problems emerge when food sovereignty, which exists as an ideal notion, intersects with diverse local mean- ings and understandings that blur the lines between mainstream and alternative food systems. For the Malian government, food sovereignty can co-exist with the state-sanctioned deterioration of peasant land rights because such actions are seen to increase domestic production. Thus the practice of large-scale irrigation through foreign investment, first instituted by the French and now vehemently opposed by *La Via Campesina*, forms part of Mali's official road to food sovereignty.

These emergent challenges around the definition and implementation of food sovereignty are not unique to Mali. Indeed, in recent years a number of questions have arisen that point to the difficulty nation states have in moving towards a food sovereignty model. These include questions about the role of international trade and the scale of alternative food systems, the location of sovereign power and the timing and extent of changes to existing power structures (see, for example, Friedmann and McNair 2008; Burnett and Murphy 2014; Edelman 2014; Hospes 2014). For Mali, one of the core challenges will be how it moves beyond the legacy of the French goal of large-scale, intensive farming for export.

Alternative food networks in the Office du Niger and the role of marginalization

This section examines the relationship between the idealized *La Via Campesina* view of food sovereignty and grounded notions of mainstream and AFNs in the

Office du Niger. It argues that historically-established gender relations shape contemporary understandings of, and participation in, mainstream and AFNs.

Defining mainstream in the Office du Niger

As with all food networks, mainstream and AFNs in the Office du Niger have emerged from historical, political and cultural processes (Jarosz 2008). As argued earlier, mainstream rice production in the region has been strongly influenced by productivist, export-oriented practices implemented by the French during the colonial period. As part of this French legacy, notions of large-scale, intensive production remain the goal of both farmers and administrators in the zone. The Malian government and the Office du Niger administration have both adopted a firmly productivist ideology to farming in the zone, which has been buoyed by successive waves of foreign aid since independence (Larder 2015). As evidenced in the national agricultural plan, intensification, mechanization and specialization remain core goals and part of the nation's vision of food sovereignty in Mali, albeit with a partial shift in focus from export to domestic markets. The desire for increased output is the state's rationale for large-scale foreign investment and genetic modification of seeds in recent years. The state argues, and many rice growers in the zone agree, that the area of irrigated lands in the Office du Niger needs to be expanded to allow for increased rice production. As I argued previously (Larder 2015), due to population growth in the 1990s there is now very limited access to irrigated land in the zone. The state suggests that foreign investment will pay for expansion of the zone's irrigation infrastructure, drastically increasing the amount of land under irrigation and thus leading to greater rice yields. Critics say foreign investors will produce irrigated rice for direct export and thus bypass national markets, which the state acknowledges will happen. However, the state argues that local farmers who produce for domestic markets will benefit from a new irrigation infrastructure such as large primary canals, which can be tapped into by secondary and tertiary canal networks.

As a result of the productivist mentality, those who grow rice are more likely to engage in high-input production that requires technologies geared towards maximizing output. For example, rice seed is more likely to be subject to scientific and technological interventions geared towards output, such as genetic engineering and selective breeding, than seed for vegetables or fruits. To illustrate, in 2008 the major farmer training and support organization in the zone, Farenfasisso, established several rice seed growers associations with funding from the Bill and Melinda Gates Foundation and Lutheran World Relief. The main objective of the associations is to improve the quality of rice seeds used by farmers in the zone, which is purportedly done by moving rice farmers away from the practice of saving seeds and towards a model of purchasing 'good' quality (genetically-engineered) seeds from the association. These associations work closely with non-governmental organizations and the Malian Government's Institute Economic Rurale to provide support in the form of buildings in which to store rice seeds, processing equipment and scales for weighing rice and training. Members of one seed association suggested that the goal of the group was to improve the quality of rice seeds used by farmers in the zone. Because of their reliance on purchased seeds, rice producers were more likely than fruit and vegetable growers to use chemical fertilizer and machines such as tractors and petrol-powered machines.

Moreover, as in the majority of farming communities around the world, the mainstream food network in the Office du Niger is a highly masculinized space (cf. Whatmore 1991; Liepins 2000; Saugeres 2002). As emphasized, this is largely a result of the way the region developed under French colonialism. The Office was developed by men with a male labour force and it was not until the 1940s that women and children began to live and work in the region. By the time women arrived men had assumed the role of producer on irrigated lands. This is not to say men's access to irrigated land is secure. Far from the radical *La Via Campesina* position that argues for producers' control of land, even mainstream rice producers in the Office have only the most precarious control of their lands. Under laws governing land access in the zone, land cannot be owned outright and access to land is granted to users on a permit system (Kater *et al.* 2000). Typically, rice farmers are granted annual leases for irrigated lands, which stipulate that tenants must pay an annual fee for using the land as well as provide labour in maintaining the canal network that supplies water to the region. Effectively, rice growers in the zone are tenant peasants, with the Office du Niger operating as a state capitalist enterprise. The Office leases irrigated rice lands to peasants for profit, which are fed back into the running, upkeep and expansion of the irrigation network in the region. These rights to land have remained unchanged since the immediate post-independence period (cf. Twagira 2014).

Defining alternative in the Office du Niger

The AFN in the Office du Niger consists of the production of fruits and vegetables on non-irrigated land for sale and subsistence. This practice is a continuation of the subsistence growing practices that first emerged in the Office during colonialism when workers started growing vegetables to supplement insufficient French-supplied rations (van Beusekom 2002). When women began to arrive during the 1940s, a division of labour emerged with men tending rice fields and women (re)assuming responsibility for food preparation (Twagira 2013). In Mali, a meal is constituted by the combination of a grain, typically rice but also millet, fonio or couscous, with the 'taste' element of a meal, known locally as sauce. Such sauces are eaten throughout Mali although different ethnicities have specialities traditional to their group. Sauce is made with condiments, nearly always oil, onion, tomato paste and Maggi-brand vegetable stock, combined with vegetables that may include potatoes, capsicum and tomatoes. Some sauces are made from the leaves of plants. In wealthier households, meat is commonly used in sauces, but for low-income households in the Office du Niger, meat is not reguarly consumed.

As part of their role in preparing food, women in the Office du Niger are responsible for the provision of fruits and vegetables. For many women, particularly those without excess cash with which to purchase goods from the market, growing their own is the only way to obtain fruits and vegetables. Women's subsistence production necessarily occurs on non-irrigated lands; the state restricts the use of irrigated land to monocropping and households prefer to consume a diversity of vegetables and fruits. Moreover, there is a strong cultural legacy in the Office of excluding women from leasing irrigated lands. Until 1998, men retained exclusive access to irrigated rice lands; leases for irrigated land could only be held in the name of the household head (typically a man). While 1998 policy reforms gave married women the right to lease lands in their own names, women still hold only a small percentage of land leases in the Office du Niger. An interview with an Office employee suggested that women hold around 3 per cent of all leases in the region, the majority of these by widows. Of the 120 women who participated in interviews and focus groups for this study, only two held leases for rice lands in their names. Thus socio-cultural norms that exclude women from cultivating irrigated fields have been slow to change despite official sanctions for married women's land use.

As a result of these factors, the majority of fruits and vegetables produced in the Office are grown illegitimately in 'gardens'. Gardens are non-irrigated lands worked by women for subsistence production. These lands are said to be 'for the family' because they are central to household food needs, but they remain places for women: 'the land beside *fala* [a natural waterway] is a place for women, where women grow fruit and vegetables. It's the place women use to produce their things' (interview with male village elder, 2011) (see Figure 5.3). The administration considers any tract of non-irrigated land to be 'undesignated' or 'vacant' and does not recognize these lands as having a productive capacity. Thus women's productive work is not recorded by the state in terms of tonnage or type of food produced and is, in effect, unseen. Consequentially, women's access to land is highly precarious. For example, in 2008 a large number of women lost access to garden lands as a result of a large-scale land deal, when a canal and road were developed to facilitate the construction of irrigation channels (Larder 2015). The state-sanctioned destruction of women's gardens was justified in terms of reclaiming illegally-used state land, which had always been designated for future development (interview with Office du Niger staff member, 2011).

While Northern conceptualizations of alternative networks – pertaining to organic production, intercropping and short-supply chains – are in evidence in the Office, the motivations underlying these practices contrast with AFNs in the minority world (Abrahams 2007). Women's vegetable production in the Office is primarily organic: producers make and use homemade compost rather than purchasing chemical fertilizers from the market. Unlike rice, which is typically sold in regional and national markets, fruits and vegetables are part of a localized distribution network. Households consume much of that produced while some of it is sold in local weekly markets. Fruits and vegetables also play important cultural roles in the local gift economy: it is customary to bring fruit when visiting relatives and guests are expected to eat with the family if they arrive close to mealtime. In contrast to AFNs in the North which result primarily from efforts

Figure 5.3 Groundnuts growing on non-irrigated, garden land.

Source: Author's own photo.

to localize and re-socialize food production and distribution (Jarosz 2008), in the Office du Niger, AFNs result from women's responsibility to provide food for their families. The latter have few options other than small-scale, intercropped, organic production.

Desires for the mainstream

Some in the food sovereignty movement have argued that women are more sceptical of industrial farming techniques than men and are therefore more likely to participate in radical movements like food sovereignty (see, for example, Wiebe, cited by Desmarais 2004). Others have suggested that women are disproportionately represented in AFNs because of their role in household food provisioning (Som Castellano 2015). In the case of women's involvement in alternative networks in the Office du Niger, such practices have formed through a combination of exclusion from access to irrigated land and the need to provide food for their families. In other words, these alternative networks are dominated by women whose exclusion from the mainstream has forced them into the realm of alterity, a finding that echoes that of Abrahams (2007).

There was no discernible difference in the ways men and women spoke about the need and desire for better access to farming inputs; both expressed a desire for better access to machines, cheaper (chemical) fertilizers, improved seeds and

Figure 5.4 Farafine nogo/organic fertilizer.

Source: Author's own photo.

training in farming methods. People in the Office distinguished between two types of fertilizers: *farafine nogo* (literally 'black input'), which referred to organic material, composed of household and animal waste products that were typically gathered in mounds in the street outside the family compound (see Figure 5.4), and *toubabou nogo* (literally 'white input'), a chemical fertilizer purchased at the markets. All interviewees saw white/chemical fertilizer as superior to black/ organic fertilizer. Women's practice of small-scale production with hand-held tools was therefore the result of financial rather than ideological barriers to large-scale farming. The inability to access land played a role, but so too did the low profitability of vegetables, which meant that it was impossible to purchase or hire machines such as tractors, pumps or even hoses to assist in farming. As a result, vegetable production in market gardens was undertaken with only a hand-held hoe. Women suggested that their lack of access to equipment severely hampered their ability to produce food for their families. As one woman said:

> You can use your strength, but this is not enough. You want to do a *great* [large-scale] thing, you need some equipment . . . with your force maybe you can feed your family, but beyond this, you can't do other things with this. Sometimes you don't even have this possibility to feed the family with your force [*sic*]. So we need the equipment.
>
> (Interview with woman in her fifties, 2011)

Women also suggested that *ingrains* (chemical fertilizers) were central to 'good' farming:

> If you want to have good crops, you have to invest much . . . all the peasants know this. Even if you have a millet field, you are obliged to put the fertilizers there in order to increase [production]: it is a must. Otherwise you won't even be able to pay the price of water so you have to put all this in order to get a good return.
>
> (Interview with woman in her forties, 2011)

Women's desire for improved access to farming knowledge and technology is underpinned by financial insecurity, but also by Malian notions of the 'good farmer'. The idea that oil-fuelled machines and chemical inputs are superior to using one's own 'force' and organic fertilizers may reflect internalization of the productivist discourse that first emerged during the colonial period. Regardless, it is clear that women want to do farming differently, with more inputs and better equipment, and that this is how 'good' farming is understood. Rather than finding value in their alternative practices, women seek to move towards the mainstream.

To this end, some women have begun producing vegetables on irrigated lands, a practice the state has sanctioned since 2009, in an attempt to increase the productivity of the region (interview with Office employee, 2011). Among the first women to grow vegetables on irrigated lands were members of Djigisemse, a women's growing collective formed in 2007. Many of the group's members lost access to their gardens as part of the 2008 land deal mentioned above on p. 114. After this loss, the group were determined to 'progress together' in order to reduce the food insecurity of their households, gain access to credit and increase their access to land for food production. In 2010 the women collectively approached the Office administration to access irrigated land for onion production. In 2011 the women planted their first onion crops in irrigated fields and reported a successful harvest. Access to this land is still insecure; women do not hold a formal lease over the land and have to negotiate with the leaseholder for permission to use the land in the off-season. Women also require access to a plot of land to sow seeds prior to planting, as their designated area for planting is still in use at the end of the rice harvest (see Figure 5.5). At the time of research, the women were reasonably confident that their efforts had been successful.

Practices in the Office du Niger and implications for food sovereignty

This final section reflects upon contemporary agricultural practices in the Office du Niger in light of moves in Mali towards food sovereignty. In examining peasants' rights to land – a critical element in the *La Via Campesina* version of food sovereignty – it is clear that producers in the zone are highly land-insecure. Those growing rice have some year-to-year security over their plot and can generally be assured that the state will not reclaim land, given the administration's emphasis on increased outputs from the zone. The same cannot be said for women

Figure 5.5 Women cultivating onion seedlings on borrowed market garden land prior to transplantation to irrigated fields.

Source: Author's own photo.

producing fruit and vegetables, whose rights to garden land are more precarious. Given their goal of a self-sufficient Mali as outlined in the National Agriculture Law of 2006, it is striking that the administration continues to fail to recognize production on non-irrigated land. By recognizing, recording and protecting non-irrigated production in the zone, the state could instantly increase domestic food production in the Office du Niger. The counter-argument, which would likely be upheld by the state, is that the government cannot profit from non-irrigated production because it cannot charge water and leasing fees. In providing more land security to women's gardens, the state would also need to overcome the long cultural legacy of the French who elevated rice production to lofty heights while downplaying other kinds of productive practices, such as subsistence gardening.

Recent moves by some women onto irrigated plots of land signal positive changes in regards to women's land rights in the region, particularly for breaking down such cultural barriers. As an Officer du Niger employee said, in the past granting leases to women was seen as promoting a culture of individualism that threatened traditional divisions of labour. He argued that Malians saw women's independence as a threat to the strong value Malians place on the family as the primary household unit:

> In the past, women were considered like a member of the family. So in the past the work was collective . . . all the family worked together. The income, it's the chief of the family who managed that. So there wasn't the 'my' individual [land] . . . Women never had the courage to ask for a field of her own from her husband, . . . in doing that, it [was] like women want to separate, to leave the home.
>
> (Interview with Office du Niger employee, 2011)

Women's move into irrigated production is helping to overcome this cultural barrier to women's autonomy within households. Historically in the Office du Niger, decisions about how irrigated land was used and how and when planting and harvesting occurred rested with the head of the household, normally the husband, father or brother. Household heads were in charge of decisions about how the harvest was allocated between consumption and sale, thereby giving wife/wives, mothers and sisters no input into decision-making over the meeting of household food needs.

While gender equality is often emphasized in the ideal food sovereignty model (Patel 2012), perhaps it is more important that Malian women themselves are starting to suggest that autonomy over food production is important to them. Yet interviews with women indicated that they still have a long way to go. For instance, women felt unable to speak to their husbands about when to harvest, despite their household's food needs possibly being severely compromised by their husband's decisions. As one woman said: 'Sometimes my husband does not work in his field . . . we cannot force the men to work; women cannot force men to work. If you tell him to go work in his field, and he refuses, what are you going to do? You can't force him to go!' (interview with woman in her twenties, 2011). Moving onto more productive, irrigated lands freed some women from having to ask their husbands repeatedly for money. For example, one woman said:

> If I grow things, *that* will bring me money. So I grow this [onion]. Otherwise, if you don't have something [money] for yourself . . . you ask your husband

'give me this, give me this, give me this' – finally he will say, 'I am fed up with you'. So you do something for yourself.

(Interview with woman in her fifties, 2011)

There was also a suggestion that autonomy was important for women who needed to support themselves into old age. An elderly widow suggested that Mali was moving from a *farafine* (black) reality to *toubabou* (white) reality and that this was resulting in the failure of children to take care of their parents into old age: 'When they [children] have some money, they go their own way, the parents will spend money on the children and then the parents suffer. Children take their money, the elderly don't work, so they can't get money' (interview with woman in her sixties, 2011). Thus while women's movement into more profitable and higher-input irrigated production may be read as a shift towards the 'mainstream', this kind of production for subsistence and sale may also be seen as a positive step towards women's autonomy, a key prerequisite of the ideal food sovereignty model (Patel 2012).

Conclusion

This chapter has examined the existence and nature of mainstream and alternative food networks in the Office du Niger, in the context of recent moves in Mali towards a particular model of food sovereignty. In thinking about the nature of AFNs in the global South, particularly in light of the global food sovereignty movement, I conclude with the following points. First, in contexts where particular groups are marginalized and mainstream productive practices are valorized, alternative-like strategies may emerge as the only option for securing livelihoods. In the Office du Niger, AFNs resemble in many ways those in the North: they are small-scale, non-industrial/organic and are predominately the realm of women. For those working towards food sovereignty, it is tempting to use such examples to illustrate peasants' desires for a different kind of food system. Yet it is important to recognize the drivers for participation in such networks as these matter for thinking about how to achieve an ideal version of food sovereignty. The motivations that underpin AFNs also matter because they help explain the choices that peasants make as they reconceptualize and renegotiate their livelihoods. For women in the Office du Niger, a long history of marginalization from the mainstream has hampered household food provisioning and decision-making, particularly given the relationship between land insecurity and alternative production practices. Women's moves into irrigated vegetable production and their desires to take up mainstream practices, such as the use of chemical fertilizers, must be read in terms of the precariousness of their AFN. Gendered farming practices in the Office du Niger can be read as resulting from the spaces men and women occupy, and in turn are excluded from, but that nevertheless act to shape both knowledge and practice.

The second and concluding point to make is that food sovereignty proponents in Mali face a challenge in navigating a way through diverse motivations in order to devise a historically and socially relevant food sovereignty model. The ideal food sovereignty model as advanced by *La Via Campesina* demands recognition of the diversity of paths that peasants follow as they address the social challenges posed by the contemporary food system. For women involved in this study, the greatest social challenge is their exclusion from land and agricultural inputs. Women's use of irrigated plots – part of the mainstream system in the zone – has gone some way to breaking down cultural barriers and improving women's access to land. At another level, food sovereignty respects the rights of states to protect and regulate domestic agricultural production and trade in pursuit of sustainable development objectives. The Malian government is advancing a food sovereignty agenda along these lines as it aims to secure domestic food needs based on national agricultural production. However, for the state this necessitates increasing national yields in part through foreign investment for expansion of irrigated lands in the Office du Niger – a move referred to as 'land grabs' by Malian activists and *La Via Campesina*. These disjunctures between the idealized version of food sovereignty and the way in which food sovereignty is being enacted on the ground point to ongoing challenges for the movement in Mali: how does the movement advance both a critical stance towards the mainstream food system as well as end violence towards women, which necessarily includes equal access to land and 'modern' inputs? What role should foreign investment play in delivering domestic food security? Such questions must be answered if food sovereignty is to come to fruition in Mali.

References

Abrahams, C. (2007) 'Globally useful conceptions of alternative food networks in the developing South: The case of Johannesburg's urban food supply system'. In D. Maye, L. Holloway and M. Kneafsey (eds) *Alternative Food Geographies: Representations and Practices*. Amsterdam: Elsevier, 95–114.

Aldrich, R. (1996) *Greater France: A History of French Overseas Exploration*. Basingstoke and New York: Macmillan and St Martin's Press.

Andrée, P., J. Ayres, M. J. Bosia and M-J. Massicotte (2014) 'Food sovereignty and globalization: Lines of inquiry'. In P. Andrée, J. Ayres, M. J. Bosia and M-J. Massicotte (eds) *Globalization and Food Sovereignty: Global and Local Change in the New Politics of Food*. Toronto: University of Toronto Press, 23–52.

Australian Food Sovereignty Alliance (2013) 'The people's food plan: A common sense approach to a tasty, resilient and fair food system for all Australians: Policy directions'. www.australianfoodsovereigntyalliance.org/wp-content/uploads/2012/10/PFP-policy-august13.pdf (accessed 27 November 2015).

Becker, L. C. (2001) 'Seeing green in Mali's woods: Colonial legacy, forest use, and local control'. *Annals of the Association of American Geographers*, 91(3): 504–26.

Benjaminsen, T. A. (1997) 'Natural resource management, paradigm shifts, and the decentralization reform in Mali'. *Human Ecology*, 25(1), 121–42.

Benjaminsen, T. A., J. B. Anue and D. Sidibé (2010) 'A critical political ecology of cotton and soil fertility in Mali'. *Geoforum*, 41(4): 647–56.

Borras, S. M. Jr. and J. Franco (2012) 'Global land grabbing and trajectories of agrarian change: A preliminary analysis'. *Journal of Agrarian Change*, 12(1): 34–59.

Boyer, J. (2010) 'Food security, food sovereignty, and local challenges for transnational agrarian movements: The Honduras case'. *Journal of Peasant Studies*, 37(2): 319–51

Burnett, K. and S. Murphy (2014) 'What place for international trade in food sovereignty?'. *Journal of Peasant Studies*, 41(6): 1064–84.

Church, R. J. H. (1951) 'Irrigation in the inland Niger delta of the French Sudan'. *The Geographical Journal*, 117(2): 218–20.

Cooper, F. and R. Packard (1997) 'Introduction'. In F. Cooper and R. Packard (eds) *International Development and the Social Sciences: Essays on the History and Politics of Knowledge*. Berkeley: University of California Press, 1–41.

Coulibaly, Y. M., J.-F. Bélières and K. Yénizié (2006) 'Les exploitations agricoles familiales du périmètre irrigué de l'Office du Niger au Mali: évolutions et perspectives'. *Cahiers Agricultures*, 15(6): 562–9.

Desmarais, A. A. (2004) 'The Via Campesina: Peasant women on the frontiers of food sovereignty'. *Canadian Woman Studies*, 23(1): 140–5.

Diallo, A. and G. Mushinzimana (2009) 'Foreign direct investment (FDI) in land in Mali'. Eschborn: Deutsche Gesellshaft für Technishe Zusammenarbeit (GTZ) GmbH.

Doss, C. (2002) 'Men's crops? Women's crops? The gender patterns of cropping in Ghana'. *World Development*, 30(11): 1987–2000.

Edelman, M. (2014) 'Food sovereignty: Forgotten genealogies and future regulatory challenges'. *Journal of Peasant Studies*, 41(6), 959–78.

Fairbairn, M. (2012) 'Framing transformation: The counter-hegemonic potential of food sovereignty in the US context'. *Agriculture and Human Values*, 29(2): 217–30.

Filipovich, J. (2001) 'Destined to fail: Forced settlement at the *Office du Niger*, 1926–45'. *Journal of African History*, 42(2): 239–60.

Food Secure Canada (2011) 'Resetting the table: A people's food policy for Canada'. http://foodsecurecanada.org/sites/default/files/fsc-resetting-2015_web.pdf (accessed 11 November 2015).

Friedmann, H. and A. McNair (2008) 'Whose rules rule? Contested projects to certify "local production for distant consumers"'. *Journal of Agrarian Change*, 8(2–3): 408–34.

Hall, D. (2011) 'Land grabs, land control, and Southeast Asian crop booms'. *Journal of Peasant Studies*, 38(4): 837–57.

Hospes, O. (2014) 'Food sovereignty: The debate, the deadlock, and a suggested detour'. *Agriculture and Human Values*, 31(1): 119–30.

Jarosz, L. (2008) 'The city in the country: Growing alternative food networks in metropolitan areas'. *Journal of Rural Studies*, 24(3): 231–44.

——(2014) 'Comparing food security and food sovereignty discourses'. *Dialogues in Human Geography*, 4(2): 168–81.

——Kater, L., I. Dembélé and I. Dicko (2000) 'The dynamics of irrigated rice farming in Mali'. *Managing Africa's Soils No. 12*. Edinburgh: IIED.

Larder, N. (2015) 'Space for pluralism? Examining the Malibya land grab'. *Journal of Peasant Studies*, 42(3–4): 839–58.

La Via Campesina (2012) 'International Conference of Peasants and Farmers: Stop Land Grabbing!' *Report and Conclusions of the Conference*. Mali, 17–19 November 2011. Jakarta: La Via Campesina Notebook No. 13.

Liepins, R. (2000) 'Making men: The construction and representation of agriculture-based masculinities in Australia and New Zealand'. *Rural Sociology*, 65(4): 605–20.

McMichael, P. (2006) 'Reframing development: Global peasant movements and the new agrarian question'. *Canadian Journal of Development Studies*, 27(4): 471–83.

Morabito, V. (1977) 'L'Office du Niger au Mali: d'hier à aujourd'hui'. *Journal des Africanistes*, 47(1): 53–81.

Oakland Institute (2011) *Understanding Land Investment Deals in Africa: Country Report: Mali*. Oakland, CA: Oakland Institute.

Osborn, E. L. (2011) *Our New Husbands Are Here: Housholds, Gender, and Politics in a West African State from the Slave Trade to Colonial Rule*. Athens: Ohio University Press.

Patel, R. (2012) 'Food sovereignty: Power, gender and the right to food'. *PloS Med*, 9(6): e1001223. doi: 10.1371/journal.pmed.1001223 (accessed 12 December 2015).

Republic of Mali (2006) Loi d'Orientation Agricole (LOA), Loi No. 06–045, 5 September 2006, C. F. R.

Saugeres, L. (2002) 'Of tractors and men: Masculinity, technology and power in a French farming community'. *Sociologial Ruralis*, 42(2): 143–59.

Som Castellano, R. L. (2015) 'Alternative food networks and food provisioning as a gendered act'. *Agriculture and Human Values*, 32(3): 461–74.

Songre, A. (1973) 'Mass emigration from Upper Volta: The facts and implications'. *International Labour Review*, 109(2–3): 209–25.

Suret-Canale, J. (1971) *French Colonialism in Tropical Africa: 1900–1945*. London: C. Hurst and Company.

Twagira, L. A. (2013) 'Women and gender at the Office du Niger (Mali): Technology, environment, and food ca. 1900–1985', doctoral dissertation, Rutgers, the State University of New Jersey, New Brunswick. http://hdl.rutgers.edu/1782.1/rucore 10001600001.ETD.000068985 (accessed 3 March 2015).

——(2014) '"Robot farmers" and cosmopolitan workers: Technological masculinity and agricultural development in the French Soudan (Mali), 1945–68'. *Gender and History*, 26(3): 459–77.

van Beusekom, M. M. (2000) 'Disjunctures in theory and practice: Making sense of change in agricultural development at the Office du Niger, 1920–1960'. *Journal of African History*, 41(1): 79–99.

——(2002) *Negotiating Development: African Farmers and Colonial Experts at the Office Du Niger, 1920–1960*. Portsmouth, NH: Heinemann.

Whatmore, S. (1991) *Home Farm: Women, Work and Family Enterprise*. London: Macmillan.

Wittman, H. (2011) 'Food sovereignty: A new rights framework for food and nature'. *Environment and Society*, 2(1): 87–105.

6 Local food, imported food, and the failures of community gardening initiatives in Nauru

Amy K. McLennan

It was a hot, dusty day in Nauru, a small island nation in the mid-Pacific Ocean. The small puffs of sea breeze were a welcome contrast to the still, sun-baked air. From where Talena and I sat, under the shade of a sheet of corrugated iron perched on four metal poles, we could see the ocean glittering in the distance. The tide was out and water lapped at the edge of the reef about thirty metres offshore. To our left was her house; large concrete bricks rose almost all the way to the roof, but stopped short. The family had run out of money during the economic downturn, before they could finish construction of the family home. Talena had agreed to share her life story with me, to help me as I sought to understand the changing Nauruan lifestyle. The high rates of lifestyle-related health concerns such as obesity, diabetes and cardiovascular disease experienced by the local population made the changing lifestyle a central interest in my research.

I completed my formal preamble and consent questions. The local community was so familiar with my research by this stage that this explanation of my interest in Nauruan history and health, particularly obesity and diabetes, seemed a bit superfluous. Before I could ask a question, Talena made a comment that had become a familiar response to my introduction: Nauruans have health problems today because people cannot access or afford the sorts of foods recommended as 'healthy', especially fruits and vegetables. She explained how, during the wealthy days following political independence in 1968, money from mining royalties, land rental payments and salaries allowed her to purchase the sorts of foods so regularly talked about by the Department of Health:

> There are exports from overseas, fruits and veggies. I used to buy vegetables! I collect[ed] all my children's salary (government salaries are collected by employees or a nominated family member on a fortnightly basis). I even spent $400 [per fortnight] on my veggies, because we really like veggies. Me, I used to come back from when I get pay, I used to have punnets of oranges and pears and apples and . . . . My fridge is even full of veggies [which] has come from overseas.
>
> (Talena)

However, the good times did not last. Corruption, lack of education and poor investment advice combined to nibble away at the long-term investments intended to provide for the people of Nauru once the phosphate mines were exhausted. When mining revenues slowed in the late 1980s, the economy began to turn. Salaries plummeted, and the National Bank of Nauru struggled into the 2000s. The cost of living did not change. Imported perishable foods quickly became luxury items that many could not access or afford.

> The fruit didn't come [any more], the prices have increased. We can't even afford them! I know *I* can't afford them, because I [was] the only one working in [an] office. Even [now] my kids [are] working [again], one or two or three . . . two of them working, I can't even say, 'give me some money to buy that', because they, too, need their money. If I boss them with their money, they will get tired of [frustrated with] working!
>
> (Talena)

Talena explained that she wanted to be healthy and to purchase healthy foods such as fruits and vegetables. However, as the sole breadwinner in her family, she could not afford such luxuries. Once her children began working again, she could not take their salaries as she had in the past. She recognized that her children would get 'tired' (a literal translation from a Nauruan term, *kenongenang*, which in practice is often used to mean 'frustrated' and 'sick of the situation') if they could not choose how to spend their own money. Instead, as I have elaborated elsewhere (McLennan 2013), she would purchase individually-packaged food products or take-away meals for herself, and her children did likewise: this was the fairest approach that helped maintain good family relations. Her story of no longer being able to access or afford fresh foods was familiar among those who had experienced the economic prosperity of post-war Nauru and the relative scarcity of recent decades.

Government-sponsored interventions have recently been implemented in Nauru to address the shortage of fresh produce, including programmes to encourage people to grow fruits and vegetables in home kitchen gardens or community farms. Such initiatives that encourage local food production have increased across the world over the past twenty years (Hawkes 2013). Since the 1990s, the Food and Agriculture Organization (FAO) has promoted 'micro gardens', 'urban gardens', 'school gardens' and 'home gardens' worldwide through the World Food Programme, with the objective of improving food security. Since the early 2000s, and following a shift in emphasis from nutrient- to food-based dietary guidelines (WHO 1999), the World Health Organization (WHO) has formally considered local gardens to be potential avenues for improving diet and diet-related health (WHO 2003). Community gardening has been linked to improved wellbeing (McCormack *et al.* 2010) and community empowerment and involvement, particularly in the United States (Blair *et al.* 1991; Armstrong 2000). Research from the United States has shown how gardening has contributed to strengthening social relations (Sommer *et al.* 1994), improving psychological wellbeing (McBey 1985) and increasing physical activity levels and fresh food consumption (Blair *et al.* 1991; Alaimo *et al.* 2008). It has also demonstrated that community gardening can reduce food costs (Armstrong 2000) and decrease the consumption of sweet foods and drinks (Blair *et al.* 1991).

Community gardens are framed by organizations such as the WHO and FAO as alternatives to mainstream industrial foodways insofar as they can supply communities, especially low-income communities that have long been dependent upon food imports, with fresh food that would otherwise be unaffordable or inaccessible to them. The local food supply provided by community gardens is removed from the global capitalist marketplace, where multinational brands provide cheap calories and unhealthy options. In other words, community gardens facilitate a transfer of power away from global entities and towards more 'sovereign' local communities. Evidence about community gardens in the United States is largely positive, and so it is generally assumed that similar models will be equally effective in non-Western settings. The failure described here of community garden initiatives in Nauru calls such assumptions into question. As Tsing argues with respect to the adoption of environmental movements of Western origin in Indonesia, 'unequal encounters can lead to new arrangements of culture and power' (2011: 5). While Tsing is largely optimistic about the possibilities of global connection that enable alternative movements, the example of community gardens in Nauru illustrates how new arrangements that arise through globalizing practices and knowledges are not necessarily positive.

On the tiny island nation of Nauru, household kitchen gardens and local community farms feature prominently in the National Sustainable Development Strategy 2005–2025. Such initiatives are intended to ensure food security, improve health and wellbeing, promote gender equality, strengthen community participation and contribute to economic development. Through the Department of Agriculture, in collaboration with the Department of Women, the Department of Health and the Department of Commerce, Industry and Environment, the Government of Nauru aims to increase domestic agricultural production with the goal of 70 per cent of households successfully establishing kitchen gardens by 2025 (Government of Nauru 2009). Yet the interim milestone – for just 10 per cent of households to establish kitchen gardens by 2008 – was not achieved.[1]

This chapter explores why alternatives to the industrial food system such as kitchen and community gardens not only fail to thrive in Nauru, but also become spaces for contestation and social friction. I begin by describing historical changes in food supply and dietary health that have occurred on Nauru over the past century. These changes have culminated in a diet of imported, industrially-processed foods, contributing to a significant burden of obesity and nutrition-related non-communicable diseases in Nauru since the 1960s (Taylor and Thoma 1983; Government of Nauru and WHO 2007). I then explain complex customs of land ownership in Nauru which affect current initiatives intended to increase fresh fruit and vegetable production and consumption. The first two sections on Nauru's food supply, diet and customs of land ownership provide an essential background for the subsequent ethnographic examples, which explain why local and internationally-driven food growing initiatives on the island became contested – and later abandoned – spaces. As Wilson also suggests in Chapter 7, in order to understand the successes and failures of a nation's efforts to produce food for its citizens, it is necessary to interrogate the politics and social relations underpinning (post)colonial patterns of land use, food supply and demand.

This chapter is based upon ten months of ethnographic fieldwork in Nauru conducted between 2010 and 2011. Data collection predominantly consisted of participant observation and life history interviews (over fifty in total, with Nauruan people aged from twenty to eighty-three years old; pseudonyms are used to describe all interviewees). The ethnographic research was complemented by extensive archive searches in libraries in Australia, Nauru and the United Kingdom.

Historical changes in the Nauruan food supply and dietary health

The Republic of Nauru, located in Oceania, is the world's smallest nation. It is a single coral atoll measuring approximately six kilometres long and four kilometres wide. Most of the atoll is composed of high-grade phosphate covered with a thin layer of micronutrient-poor topsoil which makes it unsuitable for growing crops. Moreover, a legacy of twentieth-century phosphate mining has left a large proportion of the island's surface a rocky and uninhabitable 'moonscape'. The comparatively fertile coastal rim is densely covered in housing; the majority of the island's population of 10,000 people lives on this thin strip of land which measures less than four square kilometres in area.

Pre-colonial Nauruan island ecology was characterized by a narrow diversity of food sources, but a comparatively rich array of food products (Wedgewood 1936). Fish were, and remain, central to the Nauruan diet.[2] Two fruit trees in particular were also important: *epo*, the pandanus tree (*Pandanus tectorius* and *Pandanus pulposus*);[3] and *ini*, the coconut palm (*Cocos nucifera*).[4] There is strong evidence that many products based on these seasonal foods were made and stored for years at a time through preservation practices such as cooking, drying and sweetening, and through social norms for storage and distribution. Preservation was a collaborative event that required family and community to gather together, often for months at a time, in the otherwise uninhabited centre of the island (Kayser 1934; Wedgewood 1936; McLennan 2013).

The colonial impact of the late nineteenth century changed these foodways dramatically, giving rise to a rapid increase in reliance upon imported foods. In Europe, new trade policies, railway and shipping links, and the development of methods of mass industrial food production were linked to a growing presence of imported provisions in British colonies such as Nauru. Nauru is of particular interest in investigating intersections between local and global food networks, for Nauruan households have relied almost entirely on imported foods since this time. The historical dominance of imported foods is firmly embedded in contemporary

Nauruan foodways: white rice is considered the 'traditional staple' of Nauruan people, even though rice has never been cultivated on the island. The people of Nauru incorporated foreign foods into their lives and culture, which were changing as a direct result of colonial, religious and ecological influences. As in other colonized island economies (for example, Trinidad and Cuba, see Wilson, Chapter 7) imported foods included fruits and vegetables, polished white rice, white flour, refined sugar and tinned preserved meat (Administration of Nauru 1922). They were tasty, convenient, linked to social status and offered, as colonial health authorities emphasized, much-needed dietary variety. The foods were sold in the European supply store on Nauru and their consumption was encouraged by colonial officials. They were affordable to Nauruans who earned money via salaried employment, land-related income derived from mining or other purposes (such as land rental for expatriate accommodation or government offices) or through trading *copra*[5] or fish.

By the mid-twentieth century, coconuts and pandanus, along with the skills and knowledge required to process them into a variety of preserved food products, had been largely forgotten (Wedgewood 1936). The Japanese occupation of Nauru during the Second World War starkly highlighted the island's almost complete dependence upon imported food. As supply ships were frequently torpedoed, the island's large population of Japanese soldiers and Nauruan people survived predominantly on Japanese-farmed pumpkins (which had to be fertilized using large amounts of human excrement), rice rations and fish (Government of Nauru 1994). Regular food supplies resumed after the War, when colonial authorities intensified phosphate mining activities. These were accompanied by aggressive policies for Nauruan population growth, as the War had led to a significant population decline. At the same time, the people of Nauru began to fight for greater financial compensation for their resources, more autonomy and, eventually, political independence.

Political independence in 1968 gave the people of Nauru control over phosphate mining. Mining revenues escalated in the 1970s as phosphate began to be sold at world market prices rather than at cost-price to colonial authorities. The population continued to increase, resulting in the rapid urbanization of the coastal rim, as well as housing shortages. Political independence and associated social, political, economic and legal changes, combined with increased mining-related wealth, both limited Nauruans' access to land and permitted them to freely import goods from all over the world. Electricity was introduced in the 1960s, and refrigerators and freezers meant that fresh foods could be chilled for transport to Nauru for retail or private use. Retail stores and restaurants, particularly those operated by Chinese traders, expanded across the island. Eating in local Chinese restaurants and consuming individually-packaged, processed food when busy lives permitted became increasingly common (McLennan 2013).

The economy slowed during the 1990s and collapsed in the early 2000s. In many cases, salaries fell and even stopped. While historical household incomes are difficult to estimate, they broadly fell from several thousand Australian dollars per month to several hundred (where AU$100 is equivalent to approximately GB£50 or US$75). Yet imported food prices remained constant. Imported perishable foods are much more expensive to transport and store than industrially-processed alternatives, and so fresh foods became a comparatively rare and valuable commodity. Whereas a can of ship-imported corned beef might cost fifty cents or one dollar, a bunch of Chinese cabbage grown locally, or several imported (wizened) carrots, would cost in excess of five dollars, if available at all.

Despite a long legacy of public health research and intervention (for example, Taylor and Thoma 1983; Government of Nauru and WHO 2007), and an even longer history of recommendations to alter food imports in Nauru,[6] the rate of increase of obesity in Nauru between 1980 and 2008 was four times higher than the mean global increase (Finucane *et al.* 2011). Today, Nauru has some of the highest levels of obesity, diabetes and cardiovascular diseases in the world. As a result, life expectancy has remained stagnant since the 1960s, at approximately fifty years for men and sixty years for women (Carter *et al.* 2011). In the early 2000s, poor dietary health led to a mounting emphasis on kitchen gardens and community farms in Nauru. Local food-growing initiatives were driven by donor agendas focusing on health (increasing the consumption of fresh fruits and vegetables), economic development (increasing household incomes by increasing the sale of fresh fruits and vegetables) and community empowerment (especially by increasing the participation of women and young people). Kitchen gardens were included in Nauru's National Sustainable Development Strategy 2005–2025, and they remain on the development agenda. Aid organizations have concentrated on improving awareness of the necessity for local, fresh produce for health reasons as well as income generation; they regularly provide tools and materials for developing gardens, and prizes to reward the best ones. A focus on these initiatives in Nauru illustrates the extent to which the official promotion of fresh foods and local networks of production, distribution and preparation are contested and largely unsuccessful in achieving intended health, social or economic gains.

Historical analysis illuminates the complex colonial and postcolonial under-pinnings of the contemporary food supply and nutritional health in Nauru. The following section outlines how political, economic and social relations of land ownership and land use add to this complexity. These historical conditions and relations are central to understanding the failures of projects for local food production in Nauru, which are described in ethnographic detail later in the chapter.

Politics and social relations of land ownership

Issues of land ownership are central to understanding land use, food cultivation and distribution in Nauru. The whole island is divided into irregularly-shaped and individually-named and numbered plots of land, usually of less than one hectare. In the pre-colonial period, valuable land resources were considered to be underground wells, fruit-bearing trees, building materials (predominantly coconut, pandanus and *iyo* – also known as tomano or hard-wooded beach mahogany,

Calophyllum inophyllum) and ponds for rearing *ibiya* milkfish (*Chanos chanos*). When a person owned the land, they also owned whatever resources were on (or in) it. The exception was fruit-bearing trees, which could be owned independently of the land in which they grew. Land was tended regularly and had low boundary walls (Wedgewood 1936), suggesting long-standing local customs of land division.

Land division in Nauru today is reportedly based on pre-colonial land divisions formalized by German colonial powers in the late nineteenth century in order to collect taxes and enforce curfews. In the early twentieth century a land ordinance was established that allowed the British Phosphate Commission to lease land for mining and related purposes, such as constructing processing plants or houses for labourers (Keke 1994). While claiming to have maintained traditional boundaries, colonial powers began to impose their own values upon Nauruan land ownership customs. For example, they did not recognize subtle Nauruan distinctions between the ownership of land and the ownership of plants grown on it; owners of a portion of land were assumed to also be owners of the food-producing plants on it, and so mining compensation was paid to them for the loss of both. This continued into British colonial rule following the First World War.

Colonial authorities made it a requirement by law that all landowners had a collective responsibility to keep their land tidy whether they lived there or not; those that did not comply were fined. Owners had to clean, sweep, rake, pick up litter, and maintain any dwellings or structures on their land. They reportedly did so with pride. People like Hana noticed that changing laws, along with changing ways in which profits were derived from their land – from food provision (Stephen 1936) to *copra* farming (Rhone 1921) to phosphate mining – led to changing attitudes towards the land itself.

HANA: It used to be pretty clean here . . . in the olden days. Not since the phosphate boom. My grandma used to say that Nauru was really clean because they had this . . . it's like a penalty thing. . . . Say I'm from here, and I had a piece of land over there, I had to clean it. So they penalized the landowners that didn't clean up. Before, they had to go and clean up . . . the family, on the weekends. . . . I think that's how people knew their boundaries, and where their land[s] are. Because they cleaned it up!

AKM: Why did people stop cleaning it up?

HANA: They became too rich! They became too . . . life [was] too easy, they don't have to live off the island any more. They just receive the money from it.

Today the government, supported by donors, offers sporadic financial incentives and competitions which reward individuals for caring for their land at particular moments in time. The changing economic value of land – from a cared-for site of food production and family gatherings to a source of revenue if rented or mined – changed people's relationship with it. Where families would once come together to care for their land, today they debate about the division of it and the distribution of money from it. Changing relations around, and with, land are amplified in contemporary economic conditions of uncertainty and insecurity. As illustrated in the ethnographic examples in the following section, these changing relations are central to the successes or failures of local food production.

While land is considered individually-owned, a system of collective ownership exists in practice. Like the division of land, this system is thought to have pre-colonial origins and to have been formalized by the colonial authorities. Today, land is generally inherited in equal shares apportioned to each of the deceased's children. The share that each child inherits is dependent upon the total number of children the deceased person has: if a mother of three children owns a one-tenth share of a portion of land, upon her death each of her three children will inherit one-third of the mother's share, or one-thirtieth of the total portion – and so on, such that increasing numbers of people own increasingly small shares of a portion of land. It is now rare that a portion of land is owned by fewer than twenty people. In many cases, hundreds of people might jointly own just one hectare of land (Keke 1994, 2010a, 2010b). This is not to say that the land is physically divided up; land boundaries do not officially change, although there is suspicion that corrupt officials may have re-drawn them in the past.

Inherited land shares determine the proportion of money earned through land rental payments, mining profits and other activities that generate income from the land. Major decisions about land use (for example, if one sibling wishes to construct a house on it) may only go ahead if the majority of owners (generally at least 75 per cent) vote in support of it. If the widow/widower of a landowner survives, they are granted lifetime ownership (LTO) of their deceased spouse's land and its profits until they pass away, at which time the land and its profits pass back to the descendants of the landowner and not to the descendants of the spouse or other members of the spouse's family. This applies equally whether the deceased is male or female.

In practice, the current application of Nauru's system of land inheritance is far from straightforward, perhaps as a combined result of mistrust and inter-kin tension relating to monetary disputes and economic insecurity, a growing population and administrative corruption or errors. People apply the principle of equal inheritance by all children to complex family trees as they attempt to reinterpret or reconfirm current ownership arrangements. Today, few people leave wills which stipulate alternative arrangements for land inheritance or which detail, for example, an arrangement where an adopted child might receive a greater land share. One reason for this is that people fear that if they prepare a will it will hasten their death. Another reason is that so many deaths in Nauru occur unexpectedly early from accidents, stroke and heart attack (for example, see Taylor and Thoma 1983) that there is simply no time to prepare a will. Occasionally, landowners might make verbal or written agreements (this occurred especially during the Second World War); many of these are currently being challenged by other family members who consider themselves disadvantaged and for whom oral accounts are considered insufficient or untrustworthy evidence. Official records are sometimes vague or unclear, and so they are informally or formally contested. Disputes about land ownership are also rooted in different interpretations of relatedness, where scientific notions of biological or 'blood' relatedness are sometimes at odds with ideas of kin relatedness linked to adoption, family divisions associated with divorce, re-marriage, having children out of wedlock and so on. This is further compounded by poor social memory, which is largely the result of low life expectancies and limited inter-generational knowledge transfer. Moreover, the Nauruan language is predominantly oral and aural rather than written, thus written records are not always available. Many also argue that land ownership records have been changed over time, either intentionally or accidentally, and so official records are often challenged. Disputes about land ownership are frequently unresolvable, and land claims cases are commonly contested in the Nauruan court. One result of these disputes is that the present ownership status of some plots of land is officially 'undetermined'.

Many land-related disagreements have their basis in emergent (neo)liberal values which emphasize economic self-interest. In such disagreements, individual economic interests become at odds with the interests of others and broader social interests. During 2010, for example, mining in one area was interrupted when the owners of an already mined portion of land refused to allow a road to pass through their land to permit access to a mine site on others' property further inland. They demanded financial compensation for the use of their land as a passage for traffic. Mining operations were suspended for days, and it was widely known that the Nauruan government was incurring mounting costs and debts to foreign shipping companies as cargo ships were kept waiting offshore to be loaded with phosphate.

Contemporary land ownership in Nauru is dynamic and complicated. It is shaped by historical forces but continues to be fiercely contested in the context of current relationships. This has implications for what is grown in it and on it. There is no clear agreement on whether current land ownership structures extend to food grown on the land. In practice, only the people who grow produce on the land are considered to 'own' it, but in theory, all landowners in the family may claim rights to access food grown on their land. Such competing values have implications for home and community gardening initiatives intended to address nutritional health, to which I now turn.

Ethnographic examples of local food production

Local food growing initiatives emphasized in Nauru's National Sustainable Development Strategy 2005–2025 aim to encourage people to grow fresh fruits and vegetables locally, on their own land. It is suggested that fruit and vegetable consumption, and so both food security and nutritional health, might be improved if people grow their own produce. Such initiatives may also contribute to food sovereignty and ecological sustainability, although this is not explicitly acknowledged. The sale of fresh produce should also contribute to economic security. Talena (cited on p. 127) was one of many people I met who had tried to establish a kitchen garden. Hers had been more successful than many; for a short while it had even been used by the Government of Nauru as a showcase to which donor representatives and development consultants would be brought to demonstrate the potential for successful investment. As we talked, she glanced at the rocky, dusty space of earth nearby, remembering what it had looked like as a garden and how she had had to climb up a ladder to tend to the beans: 'Yeah, in my kitchen garden I [used to] grow cabbages, beans, long beans, snake beans . . . cucumbers, um . . . and Chinese cabbage and . . . I . . . [had] some pumpkins there . . . some water-melons' (Talena). I asked her why the garden didn't exist any more, and she sighed, explaining matter-of-factly:

> [The plants] die[d] when I got tired [frustrated] (laughs) because I don't want to do [it] . . . it [was] only me and my husband doing the digging, so . . . yeah. I just come and pick them and I share to my family, brothers, sisters. I share to the community. And because I used to call it . . . um . . . the 'community's kitchen garden'. But no one comes to help in it. Only me, my cousin who's living down in the corner of the bit there . . . It's me, herself and her husband, we're doing it and then, when we collect, we harvest, we feed *all* the people. And it's not fair, like we call it the kitchen . . . the community kitchen garden, but nobody comes to turn up to help.
>
> (Talena)

Frustration with unfairness is not unique to Talena's experience. Other gardens I watched being carefully planted and cultivated failed for similar reasons. When the produce was ripe, people who had not been involved in tending the garden arrived to take their rightful share of the produce. Some gardeners, like Talena, felt a strong obligation to share with the community; a value much more common among older Nauruans I met and which is reflected in other ethnographic accounts on concepts of sharing in Nauru (Wedgewood 1936; McLennan and Ulijaszek 2014).

Reasons for a claim to garden produce varied. Some insisted that they had a right to a share because they were landowners. Others claimed a right to the produce as extended family members; sharing amongst family members is commonly practised in many Pacific island communities. Others cited past debts. Of those who did get fresh produce, some would take it home to eat but others would sell it to Chinese traders or foreigners, who paid good money for fresh produce. They could then use the money as they wished; often to purchase (imported) food as well as other goods. While in the past, status was attributed to giving fresh food to others in the community (Wedgewood 1936; McLennan and Ulijaszek 2014), now status is attached to brand-name foods, electronic equipment, clothes and cars (McLennan 2013). At the same time, these same claims left the people who had cultivated the gardens feeling angry, upset or frustrated, as if their produce had been stolen from them. As Barton explained: 'if you have a garden, then your cousin, your brother, wants his claim of vegetables . . . they don't want the other guy to succeed if they don't succeed with them . . . but they want him to do all the hard work.' There was little point in putting in hard work, he reasoned, if it was all just going to be unfairly taken by others who did not contribute.

Home gardens like Talena's are framed by donor representatives (such as those described throughout this chapter) as alternatives to the global capitalist food economy and the unpredictability and unhealthy foods which it brings. In this reading, gardens provide food sovereignty insofar as they give local people power over their own food supply by developing what the WHO (2003) calls 'small-scale [food] production'. However, as human geographers have previously argued, 'alternatives' often (necessarily) enter into wider capitalist relations and spaces (see, for example, Fuller and Jonas 2003). In the case of Nauru, while the foods produced are materially separated from global markets and the inequalities inherent in them, the local cultivation and distribution of these foods is steeped in commercial economic values of immediate and measurable material profit, loss and fairness. At the same time, donors and other external organizations are encour- aged to actively contribute to, and be involved in, this project of local food pro- duction (WHO 2003). Thus, while food is produced in a space imagined to be 'alternative' to global capitalist food networks, it is certainly not free from capital- ist systems of accumulation, appropriation and power. Instead, and echoing what Morris and FitzHerbert argue in Chapter 1 with respect to Māori potato growers, the premise that the people of Nauru live in either a capitalist or a non-capitalist system oversimplifies the complex ways in which both systems underpin food production, distribution and consumption. Cultural values of accumulation under- score Nauruans' attitudes towards their, and others', gardens, as much as values of sharing and reciprocity. As everyone feels entitled to more than it is possible to grow and share, frictions result.

Even seemingly 'successful' kitchen gardens highlight these tensions. I knew of one garden tended by a woman named Ginasii, which was thriving quite well in 2010. When I enquired about Ginasii's secret to success, I found repeated iterations of similar tensions. Ginasii's garden supplies had been largely imported by expatriate friends willing to help by either carrying goods in from overseas, or including goods like bags of potting mix in larger consignments. I also discovered that Ginasii was not very close to her extended family – they had fallen out over disputes about money and how it ought to be spent – and so she felt no obligation to share her fruits and vegetables with people beyond her nuclear family. When I asked her whether her family could help to maintain the garden in exchange for some produce, she smiled and shook her head, commenting that 'even if your family did help you, they would ask for payment'. So, while Ginasii had fewer people demanding a share of her produce, she also had fewer people to help tend the garden. This made it extremely difficult to maintain, especially given her full- time employment, and it would go through cycles of disrepair and flourishing. Ginasii's garden would frequently lapse, because when she had busy periods at work she had no time to tend her garden and water it regularly.

Similar social trends and tensions underlie the failure of community farms. In the late 2000s, the Taiwanese government, then a major aid donor to the country, supported the establishment of a national fruit and vegetable farm in Nauru. The farm was located alongside the shady Buada Lagoon in the interior of the island. It included two large plots containing fluffy Chinese cabbages, bright red tomatoes and swollen cucumbers. The two plots were separated by a dirt track. This land was selected for a range of reasons: it was in a part of the island that remained shady and vegetated, it was owned by only a few landowners and the majority of them had agreed to the lease, and the ruling political party had connections with the landowners. Personal connections make negotiating permission for lease more straightforward, while facilitating the lease of land to overseas donors channels lucrative revenues to family and friends. During my fieldwork between 2010 and 2011, I enjoyed visiting the garden every so often. Little seedlings rapidly flourished into lush vegetables as the highly-skilled Taiwanese graduate staff taught local trainees to tend neat rows of vegetables. Research was being carried out to identify the best species for the Nauruan climate, and school classes regularly visited to learn about gardening, farming and cooking. I had not visited for around four months and decided to walk past. I was surprised to see a stark contrast between the plots on either side of the track. Although one remained green and fertile, in the other the plants were withered and brown. I asked around to find out what had happened, and learned that the landowners had waited until the produce was big and then decided to raise the land rental price because they wanted to take over the farm, grow the produce themselves and either keep it for their family or sell it. So they raised the rent, which forced the Taiwanese government development workers off the land, and took over the garden. 'But', said one person, scowling, 'they [the landowners] didn't want to do the work'. They also lacked the skills and knowledge necessary to cultivate foreign plants.

The story was not quite as simple as it first appeared, however. The landowners had not simply desired to take over the farm to reap the profits for themselves. They had also taken a dislike to where the produce from their own land went. Produce from the Taiwanese farm was distributed widely; some was sold, some was given to expatriate diplomats and officials (especially the Taiwanese Ambassador), some was given to school children when they visited the farm or through a national free school meal programme, and some was given each month to Members of Parliament to distribute through their communities. I heard mixed opinions about this distribution. Some people felt it was fairly distributed. Others disagreed. But all agreed that people with money or political connections had the vegetables and the power to choose who had access to them. The landowners had responded to a situation they perceived to be unfair to them, in a way that others perceived to be greedy, selfish and not community-minded. Conflicting ideas of fairness, entitlement and value meant that, where this community farm was concerned, no one could agree on what would be the fairest way to proceed.

Friction did not only arise over established farms and gardens. It also complicated attempts to establish them. A community dialogue between donors, landowners and youth group leaders about a proposed community farm project brought some of these tensions to light. The informal meeting was being held on a sunny afternoon, but the meeting room itself was cool and dark. Sheets hung over the louvred windows to keep the heat and dust out, and an air conditioner hummed in the corner. People sat around the room on moulded plastic chairs. Most had already heard about the proposal in initial consultations; this meeting was intended to start discussing details with landowners and young, unemployed people.

The project manager, a foreign volunteer, was enthusiastic about the multi-donor cross-sectoral collaborative project. She outlined the basic agreement: donors would pay money to rent land, provide equipment (such as seedlings, gardening tools and chicken feed), and pay young people to farm and manage it. The salary would be set at AU$70 per fortnight for the first three months, then AU$140 per fortnight for the first year (the standard national wage for a Nauruan person at the time was AU$150 per fortnight), then up to AU$450 per month, with increases as the young Nauruans progressed with their training. It would essentially be a national apprenticeship programme, with formally certified training in agriculture, animal husbandry, business management and related fields, with the young people involved taking on different specializations, roles and training. The produce would initially be used to supply free infant school and primary school breakfasts in all twelve districts (a donor-funded initiative that was, at the time, using imported foods); in this way the money allocated for school breakfasts would be directed away from imported foods and towards locally produced alternatives. Any excess food could be sold by the farm to begin to build a profit base. Only three of Nauru's twelve districts would participate in the pilot scheme initially; if they were successful, donors were committed to extending the project to other districts.

The young people in the room sat quietly; shy, but also silently recognizing that the power to speak in the room rested with one particular landowner and community leader. The people of Nauru are often wary of foreigners promising money. To understand this, one only has to look to their history of colonial and postcolonial exploitation and of opportunists peddling dubious investment opportunities (for example, see Connell 2006). The foreign project manager aimed to secure the commitment of one landowning family initially to trial the scheme and demonstrate to others that it could be successful. The community leader reacted aggressively towards the project manager as she presented the project on behalf of the donors. He manoeuvred his sturdy body as if to intimidate the petite foreigner, which emphasized the clear gender imbalance in the room. Why should his community be providing 'free food' to children in schools outside of his community? Why could donors not simply give him the money so that he could choose what to do with it, rather than have a foreigner manage the project? He argued that 'this sounds like you're just taking our young people for slave labour'. And he could not understand where the money would come from in the longer term. His aggressive and angry comments were focused clearly on money and power: who would get the donor money and who would get the profits?

The community leader was not representative of everyone, but his concerns appeared to be shared by many, who sat nodding quietly in agreement. The project manager, exasperated that this man seemed selfish and ignorant of the 'bigger picture' envisaged by her and the donors, became increasingly irate as she responded to his repeated questions. They were still angrily debating when some of the people around the room, especially younger people, clearly began to disengage, staring blankly at walls or flipping out their phones. Some even left, slipping outside to wait in cars. I eventually recognized that the discussion inside was going around in circles, so I followed the others outside. They broke their silence as they began to chatter. One person wanted to punch the outspoken leader. Another wondered why the donors were dealing with such an idiot if there were lots of other districts who wanted to be involved. Others remained silent and contemplative; some seemed almost resigned in their agreement with the community leader. Most young people there said that they were keen to be involved in the programme, but they also recognized that it was not their decision to make. They could see their chance being taken away from them.

While such programmes are developed in collaboration with the Department of Women and are intended to encourage the participation of both women and young people in community activities, the way in which this programme was designed and implemented challenged community cohesion and emphasized disempowerment of these groups. Women had largely not even attended the meeting, as donors dealt mainly with people recognized as leaders through colonially imposed frameworks of community leadership. Ironically, there is evidence of female community leadership in Nauru until the arrival of colonists and the installation of an all-male Council of Chiefs in the early 1900s (Williams 1971).[7] And young people, who had gone into the room feeling optimistic about the project, had departed feeling powerless and disengaged. In the end, the project was never initiated.

As with some food sovereignty movements, the establishment of kitchen and community gardens in Nauru is not a local or autonomous response to (post)-colonial or corporate foodways. Rather, gardens are bound up in regimes of local and global power. As initiatives imposed by development partners, they remain rooted in capitalist frameworks which emphasize foreign values and de-emphasize local values and networks of social relations. Over time, these foreign values become entangled with local ones, creating new and hybrid moral landscapes. Nauruan home and community gardens illustrate this entanglement, and show how important aspects of Nauruan life, such as land ownership, attributions of status and social relations, underpin the successes or failures of foreign-designed projects geared towards the production of local, healthy food. People such as Talena lose interest in gardening because they cannot see how it benefits them: frustrations arise from trying to reconcile remnants of an economy formerly based on sharing with more individualistic values and relations which inhere in introduced colonial or donor programmes. This friction does not just impact diets, but also community cohesion.

Conclusion

As Born and Purcell (2006) contend, it cannot simply be assumed that local or 'traditional' food systems are preferable to global ones. One reason for this is that food systems are fundamentally relational, and relations and values which underpin them change over time. In Nauru, the introduction of industrially-processed imported foods in the colonial era led to separations between local food producers and consumers, and the establishment of new relationships with global suppliers. These relationships were initially mediated by colonial powers, but in the post-colonial era such mediation was largely abandoned in favour of market freedoms and the transfer of power to private corporations. Today, relationships between producers and consumers at the local level cannot simply be re-established because networks of social relations and the values underpinning exchange relations and rights to land and food in Nauru have changed profoundly.

Moreover, it cannot be assumed that nutritionally healthy foods, wellbeing, economic development and community empowerment are always positively associated. Relationships between these are made even more complex by local and global social histories as well as changing social relations and values. In the Nauruan case, encouraging economic profit within the community appears to undermine efforts to improve health. Fresh food – understood from a nutritional perspective as nourishing – serves to both strengthen existing power and health inequalities and reduce community cohesion. This, in turn, can have negative health consequences (see, for example, Holt-Lunstad *et al.* 2010). Further, locals who do not (or cannot) consume fresh, local produce risk being criticized for failing to consume it, which may also have negative health consequences (see Brewis 2014).

Relations of power are central to the successes or failures of kitchen gardens and local farms in Nauru. In all examples presented, fresh, locally grown produce appears inherently unfair as it is not necessarily available to those who do the work, but is instead distributed to those in certain positions of power: those who own land, those who have the money to purchase fresh produce and/or those who have social connections to politicians or foreign aid workers and employees. Gardens fail because in local settings power relations inherent in locally grown food are transparent and experienced on a daily basis as fundamentally unfair. Talena at once sensed pride at growing a garden and then disappointed that nobody else came to help. Ginasii was happy that she could provide for her own family and enjoyed watching things grow, but was also frustrated by her extended family's constant demands. Landowners involved in local farms were often angry at the ways in which the state or donor agencies dictated food distribution, even in cases when food was distributed to some of the most vulnerable (such as through the national infant school breakfast programme). This frustration has the potential to destabilize local structures of governance.

In contrast, the relations implicit in global, imported foods are opaque and distant in Nauru. From a local perspective, these foods are more democratic; they are uniform, individually-packaged and available to all at what is perceived to be the same fair (and comparatively cheap) cost to everyone. As Talena, and many others, found, family friction was reduced if everyone purchased their own foods on their own terms. At the supermarket, fairness and equivalence could be clearly calculated and negotiated, and power relations were not experienced or felt when food came from such a long way away. In addition, the supermarket was reliable and trustworthy, unlike gardening, and so could buffer the unpredictability of gardening for people such as Ginsaii.

While injustices of the global food system are evident to the ethnographer, this global political economic view does not necessarily translate into everyday lived realities for people as they seek to feed their families, conspicuously consume certain products and maintain good relations with their community and friends. Recent efforts to introduce locally produced food focus on the resources required – seeds, water, skills and so on – but not on the producer–consumer relationships at the heart of food production, distribution and consumption. Food-based approaches to health in Nauru may be unsuccessful because in focusing on food they do not take into account long-term social ties, relations and hierarchies that structure food networks over time. In places like Nauru, the reconfiguring of relationships – between people within the community, between people and their land, between food producers and consumers and between locals and well-intentioned outsiders – underpins the failure of sovereign spaces of food production and consumption to emerge. While food once served to reinforce and strengthen these relationships, changing power relations and cultural values mean that it now creates friction instead.

Notes

1 In the 2002 census, there were 1,677 households in Nauru (living in a total of 1,652 private dwellings); 70 per cent would be approximately 1,174 households, and 10 per cent would be approximately 168 households.
2 Fish are eaten raw, smoked, sun-dried or cooked (Kayser 2003). There is no evidence of pre-colonial fermentation practices on Nauru.
3 Pandanus fruit was an important staple carbohydrate in Nauru until regular trading was established with European ships in the late 1800s. It was 'considered a more important food than coconuts, so much so that if a man possesses three acres of land he will plant two of them with pandanus' (Stephen 1936: 53). The preservation process was a community event held in the centre of the island (Kayser 1934). Two key products were *edongo*, dried pandanus-cake which looked like a strip of leather and reportedly tasted like dried figs, and *ekareba*, pandanus flour.
4 Coconut food products include the soft jelly-like flesh and sweet milky liquid of the young green fruit. Mature, brown coconuts yield sturdier flesh which can be grated and squeezed for a cream rich in fat. The spathe of the coconut, or pod that forms around a palm flower, can be tapped to collect a sweet sap called toddy. Nauruans preserved toddy through a cooking process which concentrated the sugar and resulted in a thick red-coloured syrup, *ekamwirara*. If coconuts are stored for over three years then the flesh becomes soft and yellow. In the past Nauruans ate this as a snack called *emette* (Wedgewood 1936). Coconut flesh can also be dried in the sun and pressed to extract coconut oil.
5 Rhone (1921: 561) reports that the product called *copra*, sun-dried coconut flesh used specifically for oil (fuel) extraction that was Nauru's primary export pre-1905, was not made in the Pacific before 1872.
6 George Bray's (1927) dietary survey reported that everyday foods were coconut products and fish, as well as navy biscuits and refined white sugar. Foods such as tinned meats, rice, flour and fruits were also being imported and incurred no customs duty (Administration of Nauru 1922). Bray made a series of recommendations to the colonial authorities. A toddy emulsion rich in Vitamin B was given to infants and quickly improved infant mortality rates; however, his recommendations to ban sugar and make wholemeal flour and brown rice the staple cereals of the island were not implemented.

Nancy Kirk's (1957) dietary survey found the staple foods in all Nauruan houses to be white rice, white bread, sugar, tea and tinned meat. Negligible amounts of locally sourced fresh food were consumed: only one family surveyed caught fish regularly, and this was often sold to other families or foreigners. While the recommendation to introduce powdered milk in schools was followed, the recommendations to subsidize vegetables and fruits and to increase imports of brown rice were not. Such studies highlight the historical depth of dietary change in Nauru.
7   Since colonial records began, only two women have ever held formal national leadership roles (District Chiefs in the colonial period, or Members of Parliament in the post-independence period): Ruby Thoma (née Dediya, 1980s) and Charmaine Scotty (2010s).

References

Administration of Nauru (1922) 'Report on the administration of Nauru, 17th December 1920 to 31st December 1921 (prepared by the Administrator for submission to the League of Nations)', Melbourne: Printed and published for the Government of the Commonwealth of Australia by Albert J. Mullett, Government Printer. National Library of Australia, Nq 354.968 AUS.

Alaimo, K., E. Packnett, R. A. Miles and D. J. Kruger (2008) 'Fruit and vegetable intake among urban community gardeners'. *Journal of Nutrition Education and Behavior*, 40: 94–101.

Armstrong, D. (2000) 'A survey of community gardens in upstate New York: implications for health promotion and community development'. *Health and Place*, 6: 319–27.

Blair, D., C. C. Giesecke and S. Sherman (1991) 'A dietary, social and economic evaluation of the Philadelphia urban gardening project'. *Journal of Nutrition Education*, 23(4): 161–7.

Born, B. and M. Purcell (2006) 'Avoiding the local trap: scale and food systems in planning research'. *Journal of Planning Education and Research*, 26: 195–207.

Bray, G. W. (1927) 'Dietetic deficiencies and their relationship to disease: with special reference to vitamin B deficiency in the feeding of infants'. Australian Medical Pamphlets. Sydney: Australasian Medical Publishing Company.

Brewis, A. A. (2014) 'Stigma and the perpetuation of obesity'. *Social Science and Medicine*, 118: 152–8.

Carter, K., T. Soakai, R. Taylor, I. Gadabu, C. Rao, K. Thomas and A. D. Lopez (2011) 'Mortality trends and the epidemiological transition in Nauru'. *Asia-Pacific Journal of Public Health*, 23(1): 10–23.

Connell, J. (2006) 'Nauru: the first failed Pacific State?' *The Round Table*, 95(383): 47–63.

Finucane, M. M., G. A. Stevens, M. J. Cowan, G. Danaei, J. K. Lin, C. J. Paciorek, G. M. Singh, H. R. Gutierrez, Y. Lu, A. N. Bahalim, F. Farzadfar, L. M. Riley, M. Ezzati and the Global Burden of Metabolic Risk Factors of Chronic Diseases Collaborating Group (2011) 'National, regional, and global trends in body-mass index since 1980: systematic analysis of health examination surveys and epidemiological studies with 960 country-years and 9.1 million participants'. *Lancet*, 377(9765): 557–67. Available at www.ncbi.nlm.nih.gov/pubmed/21295846 (accessed 7 November 2013).

Fuller, D. and A. E. G. Jonas (2003) 'Alternative financial spaces'. In R. Lee, A. Leyshon and C. C. Williams (eds) *Alternative Economic Spaces*. London: SAGE Publications, 55–73.

Government of Nauru (1994) Paper submitted by the Nauru delegation at the International Forum on War Repatriations (Tokyo, 13–15 August). Nauru: Government of Nauru.

——(2009) Republic of Nauru National Sustainable Development Strategy 2005–2025 (as revised 2009), Yaren. Available at www.adb.org/sites/default/files/linked-documents/cobp-nau-2016-2018-nsds.pdf (accessed 6 June 2016).

Government of Nauru and WHO (World Health Organization) (2007) 'Nauru NCD Risk Factors STEPS Report'. Suva: World Health Organization Western Pacific Region (WHOWPR).

Hawkes, C. (2013) 'Promoting healthy diets through nutrition education and changes in the food environment: an international review of actions and their effectiveness'. Background paper for the International Conference on Nutrition (ICN2), Rome. Available at www.fao.org/docrep/017/i3235e/i3235e.pdf (accessed 6 June 2016).

Holt-Lunstad, J., T. B. Smith and J. B. Layton (2010) 'Social relationships and mortality risk: a meta-analytic review'. *PLoS medicine*, 7(7): p.e1000316. Available at www.pubmedcentral.nih.gov/articlerender.fcgi?artid=2910600&tool=pmcentrez&rendertype=abstract (accessed 29 July 2011).

Kayser, P. A. (1934) 'Der Pandanus auf Nauru'. *Anthropos*, 29: 775–91.

——(2003) *Nauru One Hundred Years Ago: 2. Fishing*. Nauru/Suva: University of the South Pacific Centre/Institute of Pacific Studies, University of the South Pacific.

Keke, L. D. (1994) 'Nauru: issues and problems'. In R. Crocombe and M. Meleisea (eds) *Land Issues in the Pacific*. Christchurch/Suva: Macmillan Brown Centre for Pacific Studies, University of Canterbury/Suva Institute of Pacific Studies, University of the South Pacific, 229–34.

——(2010a) 'Review of land tenure in Nauru: challenges to government'. Paper presented at the conference of the Australian Association for the Advancement of Pacific Studies, Melbourne, 8 April.

——(2010b) 'Review of land tenure in Nauru: challenges to government'. Draft report. Nauru: Government of Nauru.

Kirk, N. E. (1957) 'Dietary survey of the U.N. Trust Territory of Nauru'. Canberra: Nutrition Section, Commonwealth Department of Health (at the request of the Department of Territories).

McBey, M. A. (1985) 'The therapeutic aspects of gardens and gardening: an aspect of total patient care'. *Journal of Advanced Nursing*, 10: 591–5.

McCormack, L. A., M. N. Laska, N. I. Larson and M. Story (2010) 'Review of the nutritional implications of farmers' markets and community gardens: a call for evaluation and research efforts'. *Journal of the American Dietetic Association*, 110(3): 399–408.

McLennan, A. K. (2013) 'An ethnographic investigation of lifestyle change, living for the moment, and obesity emergence in Nauru'. DPhil thesis, School of Anthropology and Museum Ethnography, University of Oxford.

McLennan, A. K. and S. J. Ulijaszek (2014) 'Obesity emergence in the Pacific islands: why understanding colonial history and social change is important'. *Public Health Nutrition*. Available at www.ncbi.nlm.nih.gov/pubmed/25166024 (accessed 29 August 2014).

Rhone, R. D. (1921) 'Nauru, the richest island in the south seas'. *The National Geographic Magazine*, 40: 559–89.

Sommer, R., F. Learey, J. Summit and M. Tirrell (1994) 'The social benefits of resident involvement in tree planting'. *Journal of Arboriculture*, 20: 170–5.

Stephen, E. M. H. (1936) 'Notes on Nauru'. *Oceania*, 7: 34–63.

Taylor, R. and K. Thoma (1983) *Nauruan Mortality 1976–1981 and a Review of Previous Mortality Data*. Noumea: South Pacific Commission.

Tsing, A. L. (2011) *Friction: An Ethnography of Global Connection.* Princeton, NJ: Princeton University Press.

Wedgewood, C. H. (1936) 'Report on research work in Nauru Island, Central Pacific (Part 2)'. *Oceania,* 7(1): 1–33.

Williams, M. (1971) *Three Islands: Commemorating the Fiftieth Anniversary of the British Phosphate Commissioners, 1920–1970.* London: British Phosphate Commissioners.

WHO (World Health Organization) (1999) 'Development of food-based dietary guidelines for the Western Pacific region'. Manila: WHO.

——(2003) 'WHO fruit and vegetable promotion initiative – report of the meeting, Geneva, 25–27 August 2003'. Geneva: WHO.

7 Cuban exceptionalism?

A genealogy of postcolonial food networks in the Caribbean

Marisa Wilson

This chapter compares and contrasts two places whose divergent histories and geographies of entry into global industrial capitalism have opened up and closed off spaces for alternative food networks.[1] I trace this uneven moral and political economic development in a region that has been, from its very colonial beginnings, wholly embedded in industrial capitalist food networks: the Caribbean.

The Caribbean islands of Cuba and Trinidad share colonial histories and geographies as plantation societies, with similar patterns of trade and finance, rural and urban infrastructures, techno-scientific paradigms and social differentiations based on race, class and rural/urban location. At least until the mid-twentieth century, these social patterns swayed knowledges and values, policies and practices related to food and agriculture in both places. Food and the land on which it was grown were valued primarily for profit, in line with colonial priorities and prevailing hegemonies of techno-science and governance. But from the early 1960s the Cuban state began to disentangle its agri-food economy from these global market assemblages, which had significant effects on nutritional health and collective action around food and agriculture. Cuba's countervailing, but still industrial, food and agriculture strategy changed again after 1989, when Cubans experienced extreme food scarcities due to plummeting trade and finance in food and oil previously supplied by the Soviet Union. During this period, some parts of the state and some members of civil society converged around new and old ideas of Cuban sovereignty, including growing concerns over environmental sustainability. By contrast, food and agriculture policies in post-1980s Trinidad (and Tobago)[2] continued along the path of market liberalization and industrialization, with similar socio-economic, environmental and dietary effects as those witnessed in other postcolonial developing countries (see Chapter 6, this volume).

This chapter seeks to show how Cuban 'exceptionalism' (Whitehead and Hoffman 2007) in the domestic[3] food and agriculture sector emerged from the early 1960s, and flourished as collective action geared towards sustainable, healthy foodways after 1989, while a so-called 'plantation legacy' (Best and Polanyi Levitt 2009) based upon highly-industrialized, commodified and unhealthy food networks continued to develop on islands such as Trinidad. It does so with comparative historical research and ethnographic and policy analysis of more contemporary agri-food networks in Trinidad (Part I) and Cuba (Part II).

The ethnographic research for this chapter was conducted in Cuba between 2005 and 2007, and again during the summer of 2011, while in Trinidad the research was conducted from 2009 to 2013 (when I lived and worked in Trinidad and Tobago) and especially during the summer months of 2014. During most of these periods I lived with families as a member of their household, which provided at least partial understandings of the socialities and materialities of everyday life. I also carried out formal and informal interviews with key people driving agri-food networks in each country, including farmers, state and market actors and consumers.

Trinidad and Cuba: histories of convergence and divergence

Any social scientific study of the Caribbean must take into account the region's unique encounter with European colonialism for, as anthropologist Michel-Rolph Trouillot (1992: 22) argues, 'the Caribbean is nothing but contact'. Theorists of the Caribbean Dependency School,[4] such as Lloyd Best and Kari Polanyi Levitt (Karl Polanyi's daughter), suggest that Caribbean islands offer exceptional case studies for those interested in (post)coloniality since these places were shaped by external values and interests from their very beginnings, leaving inhabitants with few, if any, autochthonous socio-economic, ecological or dietary traditions (Richardson 1992: 206):

> As a European creation, the Caribbean has a very short history, has [had] completely open societ[ies] and has been dominated by *total institutions*. On top of this, we have the fact of [the islands'] small size. This combination yields the closest approximation possible to the social scientist's dream: a controlled experiment in society.
>
> (Best and Polanyi Levitt 2009: 2, my emphasis)

By 'total institutions' Best and Polanyi Levitt mean sugar plantations. From the seventeenth century onwards, the sugar industry shaped all aspects of Caribbean societies, from systems of governance, pricing and investment, to patterns of land use and resource distribution, to politics and hierarchies of material and social rewards. 'Foreign-owned and export-oriented' (Best and Polanyi Levitt 2009: xviii), sugar plantations were highly divisive: slavery and indentureship not only ruptured family and place histories but set one group of people against another according to imported and often re-appropriated hierarchies: blacker versus whiter, backward versus civilized, traditional versus modern, rural versus urban. Such categories and divisions were variably applied by different Caribbean populations in the periods preceding and following independence, and so will be important in what follows as I compare and contrast historical openings and closures for alternative food networks (and other counter-hegemonic actions) in Trinidad and Cuba.

Plantations were 'total' in shaping institutions, practices, meanings and values related to land use, agricultural work and food consumption. But they did not eradicate *all* opportunities for counter-hegemonic social, political and economic action. No such structure is so all-encompassing as to eliminate all possibilities for social, economic and political change. In other words, as Marshall Sahlins (1976: 43) argued, every culture contains the 'embryo of another order'. Anthropologists of the Caribbean have long maintained that the plantation system was met with resistance as well as accommodation (Olwig 1985), subversion as well as appropriation. One of the most renowned anthropologists of the Caribbean, Sidney Mintz, argued that Caribbean peasantries seized land for themselves in order to resist imposed forms of dependence wrought by the plantation. These were not pre-capitalist peasants, but 'reconstituted' peasants who formed rights and values for the land and forms of production and trade that co-existed with the capitalist plantation (Mintz 1974). As in Trinidad, food provisioning in Cuba was tied to particular international networks and moral economic relations that privileged certain ways of producing and consuming food. Yet neither state nor market 'systems of (food) provisioning' (Fine and Leopold 1993) eliminated long-established foodways that each island's reconstituted peasants had developed over generations.

The contrasting experiences of colonialism in Trinidad and Cuba created different openings and closures for alternatives to industrial capitalist foodways. Not all localized struggles became national struggles that would change the entire trajectory of an island's political economy, its socialities and geographies. While some local struggles for change were closed off by more influential ideologies and networks (as is the case of the Caribbean Dependency School in the face of 1980s market liberalism) (Potter *et al.* 2004: 331; Best and Polanyi Levitt 2009: xvii), other struggles, such as Cuba's 1898 War of Independence and 1959 Revolution, became 'relational achievements' (Massey 2011[2005]: 182) by uniting previously divided social groupings with collective moralities *and* the material means to carry them out.

Part I Trinidad

The emergence of a divided society

The nineteenth century is a good place to start for a comparative history of the two islands' convergence and divergence, for by this time there were already significant differences in the ways the plantation system was institutionalized and resisted on each island. Like Cuba, Trinidad was not a 'fully-fledged sugar economy' (Best and Polanyi Levitt 2009: 45) until the nineteenth century, a whole two centuries after British colonists on nearby islands in the West Indies, such as Barbados, St. Kitts and Antigua, had established the world's first sugar plantations (Beckford 1972: 19). For the British colonial administration, Trinidad was initially valued for its strategic position. Located only eleven kilometres from the South American mainland, Trinidad was a base from which Britain could ship its manufactures directly to Venezuela and the rest of the continent, contravening Spanish mercantilist policies (Williams 2002[1942]: 79; see Figure 7.1). Given its early role as a trading post, Trinidad developed a large, urban-oriented middle class

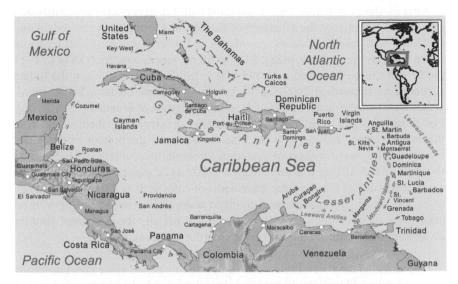

Figure 7.1 Map of the Caribbean (Trinidad and Tobago at bottom right).

Source: https://en.wikipedia.org/wiki/List_of_Caribbean_islands#/media/File:Caribbean_general_map.png.

early on, which made it different from other sugar colonies in the Caribbean (cf. Miller 1994: 20). 'By 1871, at least a quarter of the total population was urban, a high proportion for a nineteenth-century tropical colony' (Brereton 2009[1981]: 127). This urban orientation would continue into the twentieth century, leading to policies and practices that favoured exports over subsistence food production and restricted smallholders' access to land, while promoting an increasing number of industrial food imports to feed a growing urban population.

The early 'urban bias' (Lipton 1977) in Trinidad exacerbated racial, rural/urban and class divisions, as sugar work became associated with inferior categories of person. By the time sugar and slavery were introduced on a significant scale in Trinidad, there was already a racial hierarchy forming between slaves and their owners, some of whom were also of African origin. In the nineteenth century, the island was comprised of white and black merchants in urban areas and towns and white and black farmers in rural areas who grew sugar, cocoa, coffee, fruits and tubers. The latter were often referred to as 'free coloureds' (Brereton 2009[1981]: 64). Since the free coloureds had already established their place in the countryside as landowners *before* the onset of large-scale slave trading, they joined white plantation owners as slave owners in their own right: 'Slaves [in Trinidad] were much more frequently the property of coloured or even black slave-owners than in the other colonies; sometimes their owners may have [even] been relatives' (Brereton 2009[1981]: 54). Thus from the very beginnings of plantation economy in Trinidad, a racial hierarchy had been established between whites, free blacks and slaves, to which another category of person would be added after slavery was abolished in 1838: Indians.

After ex-slaves in Trinidad had completed their brutal period of apprenticeship (which was, effectively, the extension of slavery for a further six years, beginning in 1838), most 'aspiring blacks' (ex-slaves and others) sought skilled trades in urban areas (Brereton 2009[1981]: 81). To fill the labour gap, from 1844 to 1917 the Crown Colony imported over 145,000 indentured labourers from the Indian subcontinent (Sherwood 2003: 3). By the mid-twentieth century, Indian immigrants were to make up about half of the total population of Trinidad (and Tobago). For those who had spent their lives resisting or separating themselves from plantation slavery, Indians took the place of slaves at the bottom of the social hierarchy. They were 'objects of contempt for western-acculturated Africans and those of mixed African ancestry' (Singh 2002: 447), especially those in urban areas. It was during this time that inhabitants of cities such as Port of Spain began to be associated with modern and civilized ways of life, whilst people in rural areas were increasingly viewed as backward, especially those situated furthest away from white British values: Hindus who did not speak English.

> Race and class practically coincided during this period since non-Europeans were alleged by all the European colonial powers to be in various stages of unfitness for the exercise of political and economic management of their societies, and at best would have to undergo a long process of acculturation to European – in this case, British, standards and values. Only that minority of Africans and Indians fortunate enough to gain a secondary or tertiary education would be able to fulfil this requirement. In this respect, Africans and those of mixed-African descent had a significant head-start over Indians, since the latter did not become numerically significant as permanent settlers in the society before the last decade of the nineteenth century. The latter were also largely confined to the rural areas, forming the bulk of the plantation labour force, and over half their numbers were functionally illiterate in the English language as late as 1956.
>
> (Singh 2002: 445–6)

From the mid-nineteenth century until the pre-independence period in the mid-twentieth, this social hierarchy went largely unquestioned. Whites were at the top, comprising the 'political, social and economic elite'; blacks and coloureds made up the middle class 'distinguished by education and white collar jobs'; the creole working class of Africans and some Indians (who spoke English) was next; followed by the lowest rung on the hierarchy, the rural Indians who, 'although strong numerically, were separated from the rest of the population by culture and religion, by race and legal restrictions, and by their relatively late arrival' (Brereton 2009[1981]: 116).

Such urban-oriented, racialized values did not emerge in a vacuum, however, for there were concrete materialities that made some moral hierarchies more binding than others. For instance, ex-slaves who remained in the countryside found it nearly impossible to purchase Crown lands after their period of apprenticeship. After apprenticeship, the local government mandated that the smallest plot of land an ex-slave could buy was 335 acres (Brereton 2009[1981]: 89), an utterly unaffordable venture given the scant salary an ex-slave was paid during apprenticeship. A number of official and unofficial counter-hegemonic acts resisted such policies. For instance, some former slaves became reconstituted peasants who squatted on Crown lands, producing food for household subsistence and often for sale. The British Royal Commission visited Trinidad and recommended (in its reports of 1897, 1929 and 1938) that imported foodstuffs be replaced with diversified agriculture, but to the deaf ears of the local colonial government (Williams 2002[1942]: 227). Similarly, during the World Wars, a 'Grow More Food' campaign instigated by the British colonial government increased the number of acres dedicated to subsistence agriculture by nearly 10,000 acres (roughly 5 per cent of all arable land at that time). After the Wars ended, however, the campaign was abandoned (Pemberton 2002). Despite all these struggles for 'food sovereignty', the local colonial government held steadfastly to its focus on large-scale exports and food imports. Thus while the reconstituted peasants continued to produce indigenous and African crops such as cassava, sweet potato and yams, much of the food economy was based on imports.

> It's like living on a ship here. We've got twenty thousand Negroes working on the plantation from five in the morning to six in the evening, six days a week. But the scarcest thing here is food. All they do on the plantations is cocoa and cotton and sugar-cane and coffee. . . . Nearly everything we eat here is smoked or salted and comes in a box from Canada or the United States. Beef, mackerel, salmon, cod, herring. . . . Butter is orange-red with salt and costs six shillings a pound. No one thinks of churning it locally.
>
> (Naipaul 2001: 265)

By the time Crown lands became accessible to small-scale farmers in the late nineteenth and early twentieth centuries, they were mostly farmed by people of Indian origin who could purchase lands in lieu of their return passage to India (Williams 2002[1942]: 121). By then, the Trinidadian economy had begun to shift from rural-oriented sugar to urban-oriented oil, the island largely divided between black tradespeople and workers in the urban areas (some of whom were lucky enough to find jobs in the oil and gas sector) and Indian subsistence farmers and sugar workers (they were usually both) in the rural areas. Although there were some black farmers and some Indian urbanites, the government spent much energy separating the two races, especially in the countryside where there were separate latrines, schools and housing for blacks and Indians (Williams 2002[1942]: 211–12). 'Don't educate the "Coolies"', sugar planters said, 'as they will move out of agricultural labour' (Williams 2002[1942]: 212). As several historians have suggested (namely Ryan 1973; Maingot 1998; Meighoo 2003), Britain's divide-and-rule racial policy was to prove convenient for industrial capitalists in the twentieth century, since it largely precluded any unified resistance against foreign-led models for development.

Continuities in post-independence Trinidad

Although the number of counter-hegemonic acts in Trinidad increased in the periods preceding and following independence in 1962, most notably in the trade unionism of the 1930s, the anti-imperialism of the 1960s (led by the leftist campaigns of C. L. R. James) and the Caribbean Dependency School from the 1960s to the 1980s, none of these struggles was able to overcome the divisive nature of Trinidad's colonial experience. As the first prime minister of Trinidad and Tobago, Eric Williams, was to famously declare on the eve of the country's independence:

> Two races have been freed, but a society has not been formed. It takes more than national boundaries, a National Anthem however stirring, a National Coat of Arms however distinctive, a National Flag however appropriate, a National Flower however beautiful, to make a nation. The task facing the people of Trinidad and Tobago after their Independence is to create a nation out of the discordant elements and antagonistic principles and competing faiths and rival colours which have produced the amalgam that is today the approximately 875,000 people of Trinidad and Tobago.
>
> (Williams 2002[1942]: 278)

Trinidad's struggle for unity between races *and* classes remained a means by which neo-colonial interests – in the importation and promotion of industrial foods, for example – continued to shape the moralities and materialities of daily life. From the beginnings of plantation economy, class status had been associated with imported food and other metropolitan forms of consumption, which were more accessible to the (usually whiter) house slaves than the (usually blacker) field slaves (Beckford 1972: 10; Knight 1990: 121–34; Wilk 2006: 61). This attribution of symbolic capital to imported food was to influence the ways governments of newly independent Caribbean states allocated national resources and spent their national budgets (Beckford 1972: 48). Instead of developing their own agri-food economies to offset uneven terms of trade,[5] Trinidad and Tobago and most other Caribbean governments continued to depend upon food imports, foreign investments and (agri-)industrial models for development.

For instance, Eric Williams applied Sir Arthur Lewis' 'industrialization by invitation' policy in the 1960s and 1970s, which favoured foreign investments in heavy industry (oil and gas) over domestic agricultural development and opened up the country's economy to free trade and multinational corporations (MNCs) such as the US-based Texaco (which had already taken over much of the Trinidadian oil industry; Brereton 2009[1981]: 238). Caribbean Dependency theorists argue that the strong presence of MNCs in postcolonial states like Trinidad and Tobago continued to reproduce the structural inequalities of plantation economies, since externally driven networks hindered sovereign decision-making.

To the degree that the MNCs internalize financial flows, control channels of distribution, and maintain a special relationship with the metropolitan state [in this case, the United States], they are the contemporary manifestation of the merchant trading enterprises of the old mercantile era. This is the meaning of the term, 'the new mercantilism'.

(Best and Polanyi Levitt 2009: 58)

The policies of Eric Williams aligned with the US-sponsored Caribbean Basin Initiative (and earlier Marshall Plan), which encouraged Caribbean countries to join the 'free world', as opposed to the communist world, by opening up their borders to the free flow of commodities and capital (cf. Friedmann 1982: S252). His policies of the 1960s and 1970s were also compatible with popular moralities of consumption that flourished during the oil boom of the early 1970s. With a renewed confidence in the country's ability to import consumer goods with oil money, Williams assured eager urban consumers that 'money is no problem' (Wilson 2013: 112), thus reviving externally driven values for food imports that had formed since the plantation period.

Trinidadians in their seventies and eighties that were interviewed during my ethnographic fieldwork recalled the oil boom as a period of abundance but also of transition, as networks of food production and consumption continued to shift away from local geographies and towards globalized food networks, now centred on US imports:

Long ago, you would only get apples and grapes during Christmas time, and suddenly [during the oil boom] you [were] getting apples, grapes, pears . . . all the time. Suddenly they started importing lots of fruit!. . . . We used to eat more local fruits, pawpaw [papaya], pine[apple], mango. . . . Now everyone want[s] apples and grapes, all year long!

(Trinidadian in his eighties)

As this citation illustrates, during the 1970s Trinidadians became accustomed to a range of imported foods, including fruits (and vegetables). Imported and locally processed foods, too, were on the rise during this period. Part of Trinidad and Tobago's post-independence drive towards urban industrialization included development projects geared towards food canning and manufacturing, which fostered alliances between the government, local food importers and MNCs, many of which had established markets in Trinidad during the colonial period (see Wilson 2016). With government subsidies, industrialists gradually shifted from food importing in the 1960s and 1970s to food manufacturing in the 1980s and 1990s, using previously established market connections to import raw materials such as US-subsidized corn and corn derivatives (such as high fructose corn syrup, see Figure 7.2) and crude soybean oil for the manufacture of local or multinational products. Like the rise of supermarkets in the region (Reardon and Berdegué 2002), these patterns effloresced under the International Monetary

Figure 7.2 Local brands of soda in Trinidad and Tobago manufactured with high fructose corn syrup imported from the United States.

Source: Author's own photo.

Fund's (IMF) Structural Adjustment Policies, which further opened up Trinidad and Tobago's borders to increased imports of processed foods and highly-subsidized, industrially produced grains and oils (and their derivatives).

The counterpart to foreign-led urban industrialization was a techno-scientific model for agriculture that relied upon imported feed and fertilizers for the industrial production of sugar and, later, 'non-traditional' exports such as pumpkins, tomatoes and hot peppers. In 2003 Trinidad and Tobago's state-owned sugar company (Caroni [1975] Ltd.)[6] was liquidated due to decreasing profits. Though the state promised to redistribute some of the former Caroni lands to sugar workers and farmers, these promises were largely not met. Instead, the majority of state land previously dedicated to sugar (about 18,000 acres) was used for large commercial farms, industrial estates and housing developments (Richardson and Richardson-Ngwenya 2013: 10).

The demise of the sugar industry meant a drastic shift in the culture of small farming in Trinidad, and a shift away from subsistence production, since many of the former sugar cane workers and farmers had planted fruits, vegetables and tubers as well as sugar cane (Wilson and Parmasad 2014). As my Trinidadian friend told me, 'after the closure of Caroni, the people started to plant houses instead of food'. Since Indians (particularly Hindus) (Sanatan Dharma Maha Sabha 1989: 1) were most associated with rural sugar work, having been 'locked into the plantation sector by white colonial policy' (Singh 2002: 447), the shift away from sugar production in Trinidad was seen as highly politicized and racialized. During the 1990s and 2000s, many Trinidadians saw the future of sugar as a contest between rural Indians and urban blacks: 'For a government to pursue a national policy in one area only [oil and gas] and to abandon it in others is to compromise the neutrality of the state' (LaGuerre 1989, cited by Wilson and Parmasad 2014: 13).

Despite recent enclosures on their lands and livelihoods, many small farmers in Trinidad (of both Indian and African origin) continue to feed themselves and their communities, becoming part of the unseen majority of peasants across the world who feed over 70 per cent of the world's population (FAO 2013: 1). A significant amount of food continues to be produced by reconstituted peasants, but such foodways are gradually being eclipsed in Trinidad by high-calorie, industrialized diets comprised of cereal-based commodities such as high fructose corn syrup, which now may be more prevalent than sugar in contemporary Trinidadian households. And, as in an earlier period when imported food was associated with higher status, processed foods and international brands are often considered better than local 'slave foods',[7] particularly among younger generations (although from my fieldwork, it was not clear whether or how such hierarchies were racialized): 'Foreign is always better and superior. [People prefer] Heinz over Matouks [ketchup], Cadbury over Charles [chocolate], Anchor [New Zealand] over local cheese, imported sardines over local [fish], imported tins of pineapple over fresh [pineapple], apples, grapes, pears . . .' (Trinidadian in her twenties); 'Trinis believe if it's imported, it's good. . . . It's ah big brand so it is quality' (Trinidadian in her twenties); 'You know you *made it* in Trinidad when you buy your groceries at Trincity mall [which has the largest Tru Valu supermarket in Trinidad]' (Trinidadian in his twenties, my emphasis);[8] 'Anything that['s] local, that['s] not Carnival, [is] not as good' (Trinidadian in her forties).

Though the Trinidad and Tobago government now promotes local food, this is done primarily through the World Bank's (2008) model of value-added agriculture. In accordance with the model, the state's engagement with small farmers rests on the continued dominance of MNCs (including Trinidadian MNCs), for whom business-oriented farmers are seen as a key source of primary products as well as key consumers of (mostly) imported agro-chemicals (Wilson 2016). Rather than supporting subsistence farmers and agricultural diversification, then, the Trinbagonian state continues to spend large amounts of government monies on developing a single export sector (oil and gas) while importing an increasing amount of industrial food and food ingredients for local manufacturing and exporting industrially produced 'non-traditional' exports. The results are detrimental not only to environments but also to bodies. Divergence between Trinbagonian consumers who can afford to eat 'better' imported food and those who cannot, coupled with the pressure on small farmers to change their production strategies to suit the needs of MNCs, is arguably leading to the 'dual burden' of obesity and undernutrition in Trinidad,[9] a dilemma affecting many postcolonial places across the globe for similar reasons.

Part II Cuba

This section examines how Cuba and Cubans managed to disentangle themselves from highly globalized, industrial capitalist food networks stemming from colonial and postcolonial systems of value, knowledge and governance, in order to develop policies and practices geared towards socialized, healthy and sustainable food networks in the post-1990s period. Given their similar colonial histories and geographies, what kinds of material and symbolic processes enabled effective moral and political economic action in Cuba but not in Trinidad? In order to answer these questions, we must return to an earlier period in Cuban history when the sugar plantation structured the social, political economic and moral fabric of food production and consumption.

The plantation in Cuba: similarities and differences

Like nineteenth- and twentieth-century Trinidad, pre-1960s Cuba may be seen as a typical plantation economy with external controls over key sectors of the economy and powerful, if uneven, manifestations of internal complicity, often associated with race, class and rural/urban status. Yet unlike Trinidad, Cuba did not become a fully-fledged plantation society until the twentieth century. For most of the colonial period, Cuba managed to avoid many negative social consequences of the sugar plantation that had occurred in the West Indies, simply because it had long been an agrarian society centred on independent proprietors.

> Only in the late nineteenth century did the latifundium rise to destroy what had been nurtured for centuries in the process of developing a Cuban identity. The source of this destruction was the industrial concentration derived from technological change and the penetration of United States capitalism.
>
> (Higman 2002: 54–5)

The comparatively autochthonous development of the Cuban countryside from the mid-seventeenth to the nineteenth centuries was to have a profound effect on Cuban ideas of personhood and nationhood, since many Cubans had come to share a common identity as rightful owners of lands confiscated after the 'frustrated revolution' of 1898 (Wolf 1969: 253). After this so-called War of Independence, the new Republic essentially became a neo-colonial territory of the United States (Kapcia 2000: 22–3). The nature and timing of this neo-colonialism, which established a plantation economy in Cuba much later than on most Caribbean islands, is crucial for understanding how a nation-wide alternative to profit-oriented agri-food networks formed in Cuba in the twentieth century.

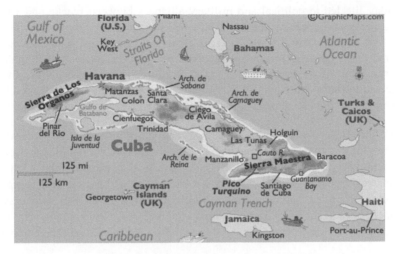

Figure 7.3 Map of Cuba, with Sierra Maestra mountain range in the southeast.

Source: http://passportandbaggage.com/cuba-factoids/; permissions granted.

As in twentieth-century Trinidad, the government of the new Republic was largely allied with the interests of US capitalists, signing the Platt Amendment as early as 1903 which, among other things, nullified rights to land previously held formally or informally by Cuba's reconstituted peasants. And, as in Trinidad, such peasants had formed in the eighteenth and nineteenth centuries alongside, or in resistance to, the slave economy (squatting on lands or escaping to undesirable areas, such as the Sierra Maestra mountains of the southeast; see Figure 7.3), although there were significant differences in their racial organization. In contrast to Trinidad, where black and 'coloured' landowners often owned and managed their own slaves, creating racial hierarchies that would only be exacerbated with the influx of Indian immigrants, farms and plantations in nineteenth-century Cuba were comprised of a mix of different social groups:

> In the atmosphere of urban, small farm and skilled slavery that prevailed in Cuba, there was no sharp break between slave and free, or between colored and white freedmen. All three groups performed the same work and often shared the same social existence in the urban centers, and in the rural areas they worked side by side in truck farming, cattle raising, tobacco growing and a host of other rural industries.
>
> (Klein 1967, cited by Wolf 1969: 4)

Moreover, intermarriage between black and white Cubans was common, as was the ability for a slave to buy him/herself free (Wolf 1969: 253).

While there was undoubtedly a social hierarchy in Cuba premised on the superiority of whiteness, which continues to the present (see, for example, Morales Domínguez 2012), early social rules of co-existence in work, marriage and everyday life led to political and military alliances between blacks and whites during the years preceding the 1898 War of Independence. During this period, Cuba was divided between wealthier loyalists to Spain, who mostly resided in Havana, and *independistas* (pro-independence fighters), many of whom lived in rural areas of the east. This unique rural/urban divide was undoubtedly connected to racial and class divisions. However, these were often eclipsed by a growing sense of national pride in the revolutionary spirit of the countryside:

> [W]hen the capital city is seen by proto-nationalist groups as the place where their interests are being betrayed, then other places, other regions, can become metonyms for the post-colonial state, a role that *Oriente* [the east] has increasingly played in Cuba over the last 150 years because of it being the setting for almost all revolutionary initiatives. In that sense *Oriente* may be thought of as the *heart* of revolutionary Cuba.
>
> (Hulme 2012: 353, original emphasis)

Hero of the 'struggle' (*lucha*) for independence, José Martí, called for Cubans of all races to join in the collective fight for a free Cuba or *Cuba Libre*. The Sierra Maestra of eastern Cuba became the focal point of the first revolutionary army, which was led by Martí and included wealthy landowners as well as reconstituted peasants, ex-slaves (called *mambisas*) and middle-class urban whites. Martí's calls for racial unity in the pursuit of national sovereignty were largely forgotten in the twentieth century, however, particularly after the United States confiscated post-war properties.

By the late 1920s, US firms controlled at least 75 per cent of the Cuban sugar industry and the majority of public utilities, mines and railroads (Leogrande and Thomas 2002: 325–6). The US 'land grab' created a massive agricultural labour force from ex-slaves and poor farmers who lost their lands, which was boosted by migrants from Haiti and Jamaica. As in Trinidad, the importation of ethnic 'Others' for plantation labour created social tensions based upon race, ethnicity and class, which were premised on an ideal 'modern' Cuban citizen (Howard 2015). US-born élites also brought in ideas of the superiority of 'whiteness', which were adopted and adapted by Cubans of all colours. Racial tensions in Cuba reached their height during the 1912 uprising in the east, when over 10,000 Afro-Cubans set fire to sugar lands and official documents, including post-1903 land titles (Pérez 1986: 533). The 1912 insurgency was brutally suppressed by the Rural Guard who killed over 3,000 Afro-Cubans, many of whom were not even implicated in the uprising (Knight 1990: 238–9). Although most Cubans abhorred these brutal tactics, the insurgency heightened racial tensions in rural areas of eastern Cuba, stifling Martí's goal of racial and, ultimately, national unity in the pursuit of independence.

In twentieth-century Cuba, many former peasants became sugar workers in field or factory, but a smaller number remained on their lands, particularly those located in the Sierra Maestra who would later become the key protagonists and symbols of Fidel Castro's 1959 Revolution. The rest of the population became proletarians and, increasingly, consumers of US products. As in Trinidad, taste preferences that began on plantations shaped the ways by which Cubans distinguished themselves from others, associating the accumulation of imported commodities with the very essence of becoming a modern and civilized person. Also similar to Trinidad, access to such valued commodities was uneven: while many urbanites could provision a range of imported luxuries such as tinned beef, white bread and sweet biscuits, people in rural areas could only procure a limited selection of basic foodstuffs (which were often imported as well) in sugar company stores:

> Though many in the city encouraged US investment and trade, developing 'new thirsts which could be slaked only with US goods' (Thomas 1971: 600), a large population in the countryside, especially in the east, was left landless and dependent on the sugar industry for both commodities (usually bought via credit at stores located in sugar complexes) and limited periods of work during the sugar harvest.
>
> (Wilson 2014a: 49)

Thus the sugar plantation and its legacies shaped material and symbolic values of 'good' food and farming in both Trinidad and Cuba. As many hankered after imported luxuries, agricultural production became increasingly tied to industrial practices based on imported agro-chemicals and machinery.

After the United States began its 'free'[10] trade with Cuba in the late nineteenth century, the majority of Cubans, particularly those in urban areas (who often owned sugar lands in the countryside) valued the kind of freedom that allowed them to access imported commodities from the new metropolis. The economic counterpart to this moral economy of free market individualism (Wilson 2014a: 11–8; 15–17; 33–7) was the 1903 Treaty of Commercial Reciprocity between the United States and Cuba, which established a bilateral trade relationship between the two countries, guaranteeing tariff exemptions on sugar in exchange for the liberalization of trade in US commodities (Pérez 1998: 29). The political counterpart was the 1903 Platt Amendment, which was much less palatable to the Cuban population. Indeed, despite the widely-held value for the freedom of the market, almost all classes in twentieth-century Cuba resented the Platt Amendment (Pérez 1995 [1988]: 244). In contrast to Trinidad and Tobago, where independence came late (in the 1960s) and without a struggle (as one Trinidadian told me, 'Trinidadians have never had anything to fight for . . . . Independence did not come with a fight, because we have always been given everything' (Wilson 2013: 114)), Cuba had already fought a collective battle for national sovereignty during its 1898 War of Independence. The Platt Amendment was a major insult to the memory of the heroes who fought in this war, many of whom were to become mythical figures in twentieth-century Cuba.

The rise of political and economic alternatives in Cuba

The Platt amendment curtailed Cuba's political freedom to act as an independent country by granting the United States unilateral rights to: (1) intervene militarily if Cuban sovereignty was threatened; (2) control all foreign policy and loans; and (3) establish military bases such as Guantanamo. By the 1920s and 1930s, the Platt Amendment became 'the focal point of growing nationalist sentiment' in Cuba (Pérez 1995[1988]: 244), manifested by counter-hegemonic ideas and acts that spanned the political spectrum, including anti-imperialist resistance movements. During this period, José Martí's ideas of Cuban unity and national sovereignty were reignited through historiography and radical organizing (most notably in the 1933 'revolution'[11]). Through these cultural and political mediums, many Cubans began to question the *moral* viability of economic freedoms offered by US commodities and wealth. Radical Cuban groups looked to Martí's writings for guidance, and words such as those cited below gave substance to growing ideologies of anti-materialism and self-determination, hard work and self-sacrifice for the *patria* (nation):

> [A]n excessive love for the North is the unwise but easily explained expression of such a lively and vehement desire for [material] progress that [Cubans] are blind to the fact that ideas, like trees, must come from deep roots and compatible soil in order to develop a firm footing and prosper, and that a newborn babe is not given the wisdom and maturity of age merely because one glues on its smooth face a moustache and a pair of side-burns. Monsters are created that way, not nations.
>
> (Martí 1975: 52)

In the above quote, values of Cuban-ness (for example, of the land) are posed as 'deep-rooted' and contrasted with the superficiality of imported values for material goods that come from 'the North' (the United States). In the citation below, Martí sets up a similar ideological contrast between the kind of materialism represented by frivolous things like 'sweet-smelling perfumes and patent boots' and the real (or 'durable') reward of hard work and sacrifice:

> Man grows with the work that comes from his hands. It is easy to see how the idle are impoverished and degraded in a few generations, until they are but fragile vessels of clay, with feeble legs which they cover with sweet-smelling perfumes and patent boots. . . . Some parents instil in the child a passion for wealth, oblivious to the fact that the only durable wealth results from the sweat of the brow.
>
> (Martí, cited by Ibarra 1986: 85, 89)

From the 1930s to the 1950s, such Martían ideas led to widespread outcries against the accumulation of wealth by corrupt politicians and those classes 'who selfishly confuse patriotism with the price of sugar' (Portell Vilá 1995[1939]: lxvii). In stark contrast to Eric Williams' 1960s mantra, 'money is no problem', predecessors to Fidel Castro's Revolution used slogans such as '*¡vergüenza contra el dinero!*' ('shame against money!') to condemn those who placed individual gain over social justice and national unity. Fidel Castro himself was to use such moral language in the late 1950s during the 26th of July Movement to overthrow President Fulgencio Batista by calling for a moralization of government and its values (Wolf 1969: 264).

Castro's speeches linked these popular sentiments with older values from Cuba's revolutionary past, such as the idea that the 'real' Cuban nation was located in the countryside and *Oriente*, where Castro and his co-conspirators set up their forces in 1958. According to subsequent myths, the 'humble' peasants of the Sierra Maestra were the first to join Castro's army, leading his eastward march towards Havana. By re-enacting the first War of Independence, the peasants and their leaders, Fidel Castro and 'Che' Guevara, became the new symbols for the revolutionary *lucha*.

In the first few years of the Revolution, Castro and Guevara adopted a radical plan that sought to overturn plantation legacies of dependency and divisiveness through policies for racial equality, agricultural diversification, price controls on basic food and household items, food rations and the nationalization of foreign-owned corporations such as United Fruit Company (who owned vast amounts of sugar lands). In stark contrast to Eric Williams' urban industrial policies, Castro and Guevara focused on the countryside, instituting two agrarian revolutions in 1959 and 1963 which redistributed foreign- and locally-owned lands to small farmers and workers (Buchmann 2009: 706). The Institute for Agrarian Reform (INRA) became 'the bastion from which the Revolution was carried forward' (Casteñeda 1998: 151) and peasants were seen as exemplars of Cuba's drive for national sovereignty, hard work and self-sacrifice.

The initial emphasis on agricultural diversification and peasant production was largely curtailed from the mid-1960s to the mid-1980s, however, as Cuba became dependent upon another superpower: the Soviet Union. Although politically opposed to the market liberalism of the United States, the Soviet model for food and agriculture was, ironically, based on the same techno-scientific paradigm as that implemented in Trinidad and Tobago and other postcolonial, developing countries: the industrial production of exports (for example, sugar or oil) in exchange for imports of 'food, feed and fertilizers' (Friedmann 2004: 5). In both Cuba and Trinidad, large expanses of land were used for high-input agricultural exports, while food and animal diets centred on mass quantities of agri-chemicals, tractors and industrially produced and processed grains, imported not from the United States but from the the the Soviet Union.

For René Dumont, French agronomist and advisor to the Cuban government, 1960s Cuba saw the 'death of the farm' (Dumont 1973[1970]: 68) as the majority of lands formerly worked by peasants became either collectivized or incorporated into massive state farms. Farmers who resisted top-down technocratic management or collectivization were deported, sanctioned on their *dossiers* or sent to labour camps, while a Marxist-Leninist approach curtailed small farmers' autonomy and designated them as profit-driven egoists (Wilson 2010). Instead of diversified, small-scale production, Cuba became a nation of massive agri-industrial farms producing monocultures (mostly sugar) for export. Between 1965 and 1985, imports of agri-chemicals to Cuba (including DDT) increased four times over, and Cuba imported more tractors than any other country in Latin America, at a peak of 79,000 in 1989 (Díaz-Brisquets and Pérez-Lopez 2000: 101–2).

The so-called 'looking-glass dynamic' of the Cold War thus shaped a parallel development of industrialized agro-ecologies and diets in both the East and West (McMichael 2006). Cuba shared a modernist techno-scientific paradigm with the so-called 'free world'. But there were two fundamental changes from postcolonial networks. The first was a structural shift away from neo-imperialist trade relations, which relied on 'plantation legacies' such as the dominance of externally-led institutions, for example, MNCs and the IMF. Caribbean Dependency theorists such as Best and Polanyi Levitt (2009: 25–6) argue that post-1959 Cuba marked a shift away from plantation patterns of trade and consumption since it severed 'traditional mercantile ties':

> [While dependency on the USSR established a] new quasi-metropolitan relationship in order to solve the problem of marketing the export staple, . . . the act of changing external metropolitan affiliation . . . assist[ed Cuba's] internal transformation. Established patterns of taste and techniques of production [were] disrupted. . . . Shortages [had to] be met by drawing on local resourcefulness. . . . On account of having localized the dominant sector [sugar] . . . [there was] greater freedom to resist [external] . . . pressure and . . . [the country] ha[d] the means to encourage and assist local enterprise.
>
> (Best and Polanyi Levitt 2009: 25–6)

The second and related shift away from postcolonial networks in Cuba was the way food and agriculture came to be valued. After Cuba established barter terms of trade with the Soviet Union, the price mechanism was no longer the sole determinant of supply and demand. A complex and top-down system of 'needs' was devised for countries within the Socialist bloc but also within Cuba itself. Rather than being fed by the free flow of commodities and capital, the Cuban citizenry was 'nourished' (Cubans use this term) by a state that was in turn supported by barter terms of trade with the Soviet Union (exchanging Cuban sugar for industrial food and oil from the Soviet Union). Moreover, instead of promoting a nation of *consumer*-citizens driven by modern luxuries and associated status hierarchies, the Cuban government attempted to create a society of *producer*-citizens directed by long-held national values that were increasingly becoming associated with socialism: hard work, self-sacrifice, asceticism, rurality (Kapcia 2000: 13–14). The new role of the Cuban citizen was to work hard to produce what the nation needed and to sacrifice individual desires for the sake of the Revolution. In the case of Fidel Castro's (unsuccessful) 1970 campaign for Cuba to produce ten million tons of sugar, this meant a willingness to engage in voluntary agricultural labour. In the case of Che Guevara's economic policies of the early 1960s (which were partly re-adopted in the late 1980s), this meant giving up luxuries for the sake of collective 'needs'.

Recalling Martí's ideas of hard work and self-sacrifice, Che Guevara wrote treatises for the Cuban population that emphasized collective morality over individual materiality, production over consumption:

> It is not a question of how many kilograms of meat are eaten or how many pretty imported things can be bought with present wages. It is rather that the individual feels greater fulfillment [in work], that he has greater inner wealth and many more responsibilities in our country, the individual knows that the glorious period in which it has fallen to him to live is one of sacrifice.
>
> (Guevara 1971: 352)

As an alternative to a market dominated by imported products from the United States, the Cuban food economy became a carefully regulated system of national supplies from Soviet and other[12] sources, provisioned in terms of top-down distinctions between luxuries and needs (Wilson 2014a: 55–63). Cubans who remained on the island were tasked with giving up luxuries and producing for the Revolution, while the state was seen as responsible for providing basic necessities to eradicate what had become highly geographical and racialized differences between 'haves' and 'have nots' in twentieth-century Cuba.

Fidel Castro's policies for redistribution (largely based on Che Guevara's moral economy of need) and his policies for racial, rural/urban and health equality led to significant changes in the socio-economic and geographical composition of postcolonial Cuba. Along with other *logros sociales* (social gains) such as dramatic increases in literacy, an increase in Afro-Cuban representatives on the Council of State[13] and rural access to medical facilities, in the first years of the Revolution the Cuban state managed to overcome one side of the 'dual burden' that was becoming increasingly evident in places like Trinidad: hunger and malnutrition (the other side of it being obesity, which is still a complex problem in Cuba) (see, e.g., Franco *et al.* 2013). Such successes remained evident in Cuba for decades, mostly as a result of finance from, and barter with, the Soviet Union. After the fall of the Soviet Union, however, a new official paradigm for food and agriculture was implemented, based upon small-scale, agro-ecological production and the 'battle' for national food sovereignty (Castro Ruz 2010), which I have called an alternative food network.

Cuba's alternative food network: official and lay perspectives on sovereign production and consumption

When ties to Soviet trade and aid faltered in the late 1980s, Cubans suffered severe and ongoing food scarcities. Between 1985 and 1993 overweight and obesity in Cuba fell by 52.3 per cent in men and 44 per cent in women (Rodríguez-Ojea Menéndez and Jiménez Acosta 2005: 117), extraordinarily affecting all population groups (Franco *et al.* 2013). Undernutrition once again became a significant problem, as evidenced in the large-scale outbreak of neuropathy, a disease associated with nutrient deficiencies (Gay *et al.* 1994, cited by

Rodríguez-Ojea Menéndez and Jiménez Acosta 2005: 115). The other end of the food provisioning system, production, was also a critical issue since without Soviet help Cuba's capacity to import oil, machinery parts and agricultural inputs for sugar cultivation fell by 74 per cent (Leogrande and Thomas 2002: 342). Estimated at three to four billion dollars per year, Soviet aid had reinforced Cuba's reliance on industrial food imports and sugar exports (Leogrande and Thomas 2002: 342); like other plantation economies, external dependency had industrialized Cuba's food economy. However, in the late 1980s the Cuban state began to shift focus to sustainable methods for agricultural diversification, while re-incorporating Che Guevara's moral economy of production and consumption, which was based upon a reciprocal contract between hard-working citizens and the redistributive state.

Partly in an attempt to fulfil its role as 'provider' for needy citizens (Wilson 2014a), the Cuban state continued to maintain strict controls over the country's food economy, re-distributing food imports (mostly basic foods such as whole chickens, wheat and rice) to social outlets such as care homes, schools and hospitals and designating most calorie-dense, processed food products as luxuries that could only be bought in Cuban Convertible Dollars (after 1993) and tourist venues. In contrast to Trinidad and Tobago and other developing countries, where Structural Adjustment Policies opened up national markets to foreign invest- ments and MNCs, few supermarkets and no fast food companies could establish retail outlets or advertise their products in Cuba (Rodríguez-Ojea Menéndez and Jiménez Acosta 2005: 116). Moreover, from the late 1980s the Cuban state invested in large-scale education campaigns and officially-sponsored radio and television programmes geared towards healthy eating practices (Rodríguez-Ojea Menéndez and Jiménez Acosta 2005: 116). The near-total exclusion of foreign influence in Cuba's *domestic* food markets created unique, population-wide nutritional patterns that are still under investigation.

In possible anticipation of the drastic fall of food and fuel imports, in 1987 the Cuban state implemented the National Food Programme which, along with Fidel Castro's policy of 'Rectification' sought to re-instill Martían and Guevarian values of hard work and self-sacrifice through volunteer labour in food production. Similar to the militant drive to produce ten million tons of sugar in 1970, the Castro government, headed by Minister of Defence Raúl Castro, called on 'all good revolutionaries' (Premat 2003: 87) to produce food for their families and communities with the slogan 'beans over bullets' (Eckstein 1994: 28) and, a few years later, national food sovereignty (Leitgeb *et al.* 2015). In his role as the head of the military, Raul Castro used the army to advocate local food production, starting from the Ministry of Defence headquarters in Havana where all spare lands were planted with edible crops.

As in El Salvador and other Latin American countries (see Millner, this volume), Cuba's history of *campesino* (peasant) resistance was drawn upon to promote an alternative model for food and agriculture. Drawing from Cuba's revolutionary value for peasants as true revolutionaries, from the mid-1990s (to the latest policy implemented in 2008) the Castro government has enacted a series of land reforms to cooperativize and/or redistribute state lands to thousands of small farmers, mostly officially recognized (and usually white) workers who agree to produce food for their families and communities. In stark contrast to Trinidad and Tobago's industrial model for small farming, Cuba's National Association of Small Farmers trains new and old farmers in agroecological farming practices while re-distributing farmers' produce to social outlets through the institution of the *Acopio* (which redistributes produce via official channels). Farmers' remaining produce may be sold at the increasing number of farmers' markets, which re-opened in 1994 having been shut down in 1986 during Castro's Rectification campaign. In 2011, the Cuban government removed restrictions on direct sales between, for example, farmers and tourist hotels, which boosted farmers' income.

Most post-1990s developments in Cuban agriculture were driven by state mandate. Yet urban agriculture developed from the grassroots upward, only later to be supported by state institutions such as the National Group for Urban and Sub-urban Agriculture, which was established in 1998 to support (and control) local farming initiatives. Between the early 1990s and 2012, 382,000 urban farms of two hectares or less were established on unused plots of land, household patios and rooftops in Havana, mostly producing organic fruits, herbs and vegetables for Havana's residents, who make up 75 per cent of the Cuban population (Leitgeb *et al.* 2015: 2–3). By 2005, about 8 per cent of all fruit and vegetable consumption in Havana came from the city's urban farmers (Rodríguez-Ojea Menéndez and Jiménez Acosta 2005: 116).

The grassroots emergence of sustainable urban agriculture in Havana and its surroundings stemmed more from necessity than any 'alternative' form of resistance to industrial capitalist food networks. Yet the combination of grassroots action and state support for urban agriculture and other forms of small-scale farming may be seen as a national alternative food network since it established enduring alliances between state and civil society groups, which scaled-up localized imperatives for healthy and sustainable food. While this alternative food network developed in the context of the household (rather than national) *lucha* for food provisioning, the 'glue' that holds it together is a national moral economy based upon new and old ideas of national sovereignty and socialist values of hard work and solidarity (Wilson 2014a). Urban and suburban agriculture in Cuba resembles what Smith and Jehlička (2013) call the 'quiet sustainability' of post-socialist households in Eastern Europe. It is also tied to a national project for food sovereignty, which has led to successful linkages between local and supra-local food networks.

My ethnographic fieldwork revealed disjunctures in the national alternative food network, however. For instance, farmers I interviewed complained about state controls such as top-down designations of appropriate seeds to use for subsidized foods, usually based on quantity rather than quality (Wilson 2014b; Leitgeb *et al.* 2015: 8–9). Yet farmers also expressed a pride in producing for, as one put it, 'the patrimony of the nation' without the need for external inputs: 'With the hard work of myself and my family, we are able to feed ourselves and our community. We do not need anything but what we create on the farm: worm

humus, organic soil. And we get bio-pesticides and bio-fertilizers from the state.'
My findings (Wilson 2014a, 2014b) resemble those of other researchers who have
studied urban agriculture in Cuba, whose work also suggests that small-scale
farmers promote a sharing economy based on an idea of sacrificing for the
Revolution: 'We *parceleros* [small farmers] benefit and those around us benefit
also. You yourself have just witnessed how many children drop by to ask for
guava fruit. I reap the benefits of my sacrifice and so do those around me' (Premat
2003: 91–2). Although citizen and state definitions of food sovereignty often
differed in terms of the scale at which it could be implemented (the household or
the nation; Leitgeb *et al.* 2015; cf. Edelman 2014: 967–8; cf. Edelman *et al.* 2014),
both official and lay discourses centred on the need to feed people through hard
work and with the use of national and farm-level resources.

Cuban food producers, particularly newly-landed farmers, demonstrated a
clear commitment to their nation's revolutionary alternative; however, the
'alterity' of Cuban consumers was much less clear-cut. After years of scarcities
and, before that, low-quality Soviet goods, many Cubans (particularly of the
younger generation) regained their desire to 'know the world' (Wilk 2013:
382) through imported foods, resembling well-to-do groups in Havana in
pre-1959 Cuba who revelled in their 'freedom' to access US markets. As
Højlund notes:

> Drinking cola seems to be a very particular sign of status among young
> people in the streets of Habana (Neri 2010), who are dreaming of a better
> future and – perhaps inspired by the many tourists and the modern ideal of
> globalization – are curious to explore the world outside Cuba.
>
> (Højlund 2014: 83)

Yet there were also instances during my fieldwork in rural Cuba (where
imported commodities and tourism are less prevalent) that reflected the importance
of collectivism and sharing over individual preferences for taste or novelty. The
most cogent example came when I helped my Cuban 'father' (with whom I had
lived, along with my Cuban 'mother', for many months) build our neighbour's
house. After the work was done, I offered to buy a bottle of Havana Club rum to
celebrate, although I knew that my Cuban father had a bottle of *gasolina* (very
cheap rum) stashed away. My Cuban father justified his preference for (what
I perceived to be) the lower-quality rum in this way:

> This rum is cheap, it comes from *Them* [the state]. Everyone can buy it.
> It is good because of this. And everyone got a bit of it at our neighbour's
> house. But I asked for a small bottle [from our neighbour] for tonight and
> worked harder. Now I know my neighbour will come and work for us when
> we need it.
>
> (Wilson 2012: 286)

For certain Cubans at least, 'lower-quality' goods are preferable when their provision is attached to socialities of sharing such as work exchanges. In contrast to Veblenian assumptions about the universal desirability of luxury goods, consumer moralities of collectivism sometimes trump the kind of market 'freedoms' that US commodities seem to bring.

Since President Obama's announcement of normalized relations between the United States and Cuba in December 2014, the lure of market freedoms are competing once again with Cuba's political freedom to maintain its sovereignty. As the Cuban state opens its economy to market channels that compete with official and local redistributive networks, it also calls into question long-held values that set collective needs over individualized patterns of supply and demand based on the price mechanism (state plans for the unification of Cuba's currency are likely to escalate this process; Morris 2015). What seems unquestionable is that, if Cuba's food economy opens to US markets in processed foods and agricultural inputs, the alternative food network that has been established over the past half-century will increasingly be called into question in everyday life. An increased availability of agro-chemicals will likely lead to more industrial agricultural practices (Wright 2009, cited by Leitgeb *et al.* 2015: 8), and lower costs of shipping foodstuffs from the United States (as opposed to Europe) may re-introduce Cuba's dependency on US food markets (Leogrande and Thomas 2002: 356, footnote 74).

While it is still unclear whether and how the moral economy of Cuban food sovereignty is shifting during the present conjecture of US-Cuban normalization, what is clear is that, compared to a place like Trinidad, Cuba and Cubans continue to maintain their 'exceptionalism' in the face of globalizing and industrializing foodways.

Conclusion

If alternative food networks are geographically uneven and inhabited in multiple ways, then so too are mainstream food networks. Shared understandings of 'mainstream' in relation to 'alternative', and possibilities for collective moral and political action based on these shared understandings, depend in part on the ways people, places and things become embedded in global capitalist assemblages. This chapter has shown how the divergent trajectories of Cuban and Trinidadian food networks led to contrasting meanings and practices of food production and consumption in the context of postcolonial state formation but also in relation to everyday lives and livelihoods. While Trinidad has followed the path of other market liberal countries in creating a country of consumer-citizens, food policies and practices in post-1959 Cuba have converged around the need to cultivate producer-citizens. Such a case for Cuban exceptionalism is highly relevant to debates about food sovereignty, which centre on the 'right to produce food' (Edelman 2014: 967).

Comparing these divergent agri-food networks also provides an opportunity to understand whether and how spaces for alternative food economies are opened up or closed off in the course of particular place histories. I have attempted to do this by focusing on specific 'moments' in Trinidadian and Cuban history that have, among other things, shaped the nature and extent of external influence in the food and agricultural policies and practices of each island. I have argued that similarities and differences in the ways the sugar plantation was institutionalized, and the timing of that institutionalization, fostered or hindered the emergence of a unified politics of resistance to global market food networks in Cuba and Trinidad.

According to Fernand Braudel,[14] wherever it went sugar destroyed 'ancient equilibriums' of demography and labour, resource use and environment. Guided by profit, sugar plantations were highly capitalistic, industrial and modern enterprises, shaping communities of practice that tied industrial agriculture to (increasingly) industrial diets, and creating modern divisions of labour that differentiated house slaves from field slaves through individualized rewards. On most islands of the Caribbean, such political and moral economic patterns, and their cultural effects, continued as plantation legacies such as the dominance of multinational corporations, dependencies on industrial development and its import requirements and hierarchies of consumer status that reproduced long-distance food networks. In Cuba, however, many of these legacies were overturned by a socialist government that continues to engage its populace by associating long-term values for national sovereignty with newer values such as collective needs (post-1960s) and the sustainable reproduction of national resources (post-1990s). Such values for food and the land on which it is grown have remained essential to Cuban socialism in the twenty-first century, with significant effects on Cuban bodies and, increasingly, environments.

Ultimately, I hope this chapter (and book) kindles debates about why and how alternative food networks emerge in some places and not others. It is clear from the extensive literature that collective actions flourish in many places of the global North, but what historical, social, moral and political economic conditions and environments open up such possibilities in postcolonial places of the global South? How do state- or market-led imaginaries of 'good' food and farming become embedded in everyday meanings and actions? What political and moral drivers enable more extensive and resilient alternatives to industrial capitalist food networks that can scale-up localized actions? These questions are not just matters for academic debate, but have profound implications for those concerned about the effects that industrial capitalism is having on social inequalities, environmental devastation and nutrition transition (Popkin 1998) in the global South. In this chapter I have taken a long view of postcolonial food networks to explore these implications for the cases of Trinidad and Cuba.

By tracing the genealogies of agri-food networks in Trinidad and Cuba I have attempted to show how official systems of food governance (whether market liberal or socialist) converge with lay meanings and practices of 'good' food production and consumption. My ethnographic research in Trinidad suggests that hegemonic values that align with global capitalist imperatives, such as industrial production, urbanization and food import consumption, have had a significant (if not all-encompassing) influence on everyday patterns of food production and consumption. In Cuba, by contrast, counter-hegemonic values for food and agriculture, geared towards local production, healthy eating and sustainability, are evidenced in official and lay contexts alike. These findings suggest that Cuba may have succeeded at the national level in what food movements in places such as the United States and the United Kingdom have only achieved at the local level: shorter food supply chains, de-industrialized agriculture and a shared moral commitment by consumers and producers to promote ethical, healthy and sustainable food.

Acknowledgements

I would like to thank the University of the West Indies and the Moray Endowment Fund of the University of Edinburgh for providing funding for research trips to Cuba (2011) and Trinidad and Tobago (2014), respectively. I would also like to thank my friends and research participants on the two islands, without whom this paper could not have been written. Finally, I am grateful to Amy McLennan and Isabelle Darmon for reviewing the paper, and to attendees at seminars hosted by the Department of Geography, University of Glasgow, the Food Studies Centre of the School of Oriental and African Studies and the Science, Technology and Innovation Studies Group of the University of Edinburgh where the paper was presented and discussed.

Notes

1 As indicated in other chapters of this volume, alternative food networks are 'diverse spaces of production, exchange and consumption of food that address needs for *healthy* food and accommodate *values of environmental protection* and *collective action* through community building between producers and consumers' (Gritzas and Kavoulakis 2015: 11, my emphases).
2 Trinidad is the larger island of the two-island state, Trinidad and Tobago. My focus here is on Trinidad, rather than the two-island state as a whole, since the colonial history of Trinidad is very different from that of Tobago.
3 In this chapter I am primarily concerned with producer–consumer networks that feed Cubans, not tourists or the export sector.
4 The School is also referred to as New World Dependency Theory.
5 Uneven terms of trade is a concept developed by dependency theorists which refers to the unequal exchange of raw material exports with inelastic (stagnant) prices for imports of industrial products (e.g. food manufactures) with elastic (increasing) prices.
6 The Trinidad and Tobago sugar industry was nationalized in 1975 when Tate & Lyle sold its holdings to Caroni (1975) Limited (Caroni is a region on Trinidad where most of the sugar lands were located). Although the company continued to produce and refine sugar through contract farming and large-scale production until 2003, its profits started to fall in the 1980s.
7 During fieldwork, I often heard local foods such as yams and other tubers referred to as 'slave foods', particularly by the younger generation.
8 The above quotes were also cited in Wilson 2013: 113.

9 According to the 'National Report for Trinidad and Tobago Civil Society's Review of the Progress Towards the Millennium Development Goals (MGD)' of 2013, www.commonwealthfoundation.com/sites/cwf/files/downloads/MDG%20 Reports%20T%26T_FINAL_1.pdf (accessed 8 December 2015), more than 20 per cent of the Trinidad and Tobago population is living below the poverty line, with 8 to 11 per cent of the population undernourished. Meanwhile, according to the Global Health Observatory data repository of the World Health Organization, http://apps.who.int/gho/data/node.main.A897C?lang=en (accessed 23 January 2016), in 2014 63.1 per cent of the population of Trinidad and Tobago was either overweight or obese.

10 It was not until 1870 that Spain abrogated its Navigation Acts and Cuba became open to more liberalized trade with the United States. The trade was not free, however, since the US subsidized key agricultural commodities such as wheat, corn, soybeans *and* sugar, protections that continue to the present.

11 In 1933, university professor, Dr Ramón Grau San Martín, led a group of radical activists, middle-class professionals, students and disgruntled members of President Gerardo Machado's army in a coup against the Machado government. Members of the 1933 'revolution' called for social justice for Cuba's working classes, including an agrarian reform to secure land titles of all peasants. In the 'government of 100 days' of Grau San Martín, many social reforms were passed, including women's right to vote, the eight-hour workday, a minimum wage for cane-cutters and the abrogation of the Platt Amendment (in 1934). The provisional government of 1933–1934 was replaced by a US-backed regime led by Fulgencio Batista, who was to become the key enemy of Fidel Castro's 26th of July Movement in the 1950s.

12 From the 1970s Cuba also traded with capitalist countries, particularly in Europe. Still, the Soviet Union supplied most basic foodstuffs for state rations and subsidized markets.

13 See www.coha.org/revolutionary-racism-in-cuba/ (accessed 23 January 2016).

14 Cited by Higman 2002: 53.

References

Beckford, George L. (1972) *Persistent Poverty: Underdevelopment in Plantation Economies of the Third World.* New York: Oxford University Press.

Best, Lloyd and Kari Polanyi Levitt (2009) *Essays on the Theory of Plantation Economy: A Historical and Institutional Approach to Caribbean Economic Development.* Kingston: University of the West Indies Press.

Brereton, Bridget (2009[1981]) *A History of Modern Trinidad 1783–1962.* Champs Fleurs: Terra Verde Resource Centre.

Buchmann, Christine (2009) 'Cuban home gardens and their role in socio-ecological resilience'. *Human Ecology* 37(6): 705–21.

Casteñeda, Jorge G. (1998) *Companero: The Life and Death of Che Guevara.* New York: Vintage Books.

Castro Ruz, Raúl (2010) 'Discurso pronunciado por el general de ejército Raúl Castro Ruz, Presidente de los Consejos de Estado de Ministros, en el quinto period ordinario de sesiones de la VII Legislatura de la Asamblea Nacional del Poder Popular'. Speech, Palacio de Convenciones, 1 August.

Díaz-Brisquets, Sergio and Jorge Pérez-Lopez (2000) *Conquering Nature: The Environmental Legacy of Socialism.* Pittsburgh, PA: University of Pittsburgh Press.

Dumont, René (1973[1970]) *Is Cuba Socialist?* London: Andre Deutsch.

Eckstein, Susan (1994) *Back from the Future: Cuba under Castro.* Princeton, NJ: Princeton University Press.

Edelman, Marc (2014) 'Food sovereignty: forgotten genealogies and future regulatory challenges'. *The Journal of Peasant Studies* 41(6): 959–78.

Edelman, Marc, Tony Weis, Amita Baviskar, Saturnino M. Borras Jr, Eric Holt-Giménez, Deniz Kandiyoi and Wendy Wolford (2014) 'Introduction: critical perspectives on food sovereignty'. *The Journal of Peasant Studies* 41(6): 911–31.

FAO (Food and Agriculture Organization of the United Nations) (2013) 'Coping with the food and agriculture challenge: smallholders' agenda'. Preparations and Outcomes of the 2012 United Nations Conference on Sustainable Development (Rio+20), by Karla D. Maass Wolfenson. www.fao.org/fileadmin/templates/nr/sustainability_pathways/docs/ Coping_with_food_and_agriculture_challenge__Smallholder_s_agenda_Final.pdf (accessed 8 December 2015).

Fine, Ben and Ellen Leopold (1993) *The World of Consumption.* London: Routledge.

Franco, M., U. Bilal, P. Orduñez, M. Benet, A. Morejón, B. Caballero, J. F. Kennelly and R. S. Cooper (2013) 'Population-wide weight loss and regain in relation to diabetes burden and cardiovascular mortality in Cuba 1980–2010: repeated cross sectional surveys and ecological comparison of secular trends'. *BMJ* 346: f1515.

Friedmann, Harriet (1982) 'The political economy of food: the rise and fall of the postwar international food order'. *American Journal of Sociology* 88 (supplement): S248–S286.

——(2004) 'Feeding the empire: pathologies of globalized agriculture'. In Leo Panitch and Colin Leys (eds) *The Empire Reloaded.* London: Merlin, 124–43.

Gritzas, Giorgos and Karolos Iosif Kavoulakis (2015) 'Diverse economies and alternative spaces: an overview of approaches and practices'. *European and Urban Regional Studies*: 1–18.

Guevara, Ernesto Che (1971) 'Man and socialism in Cuba'. In Bertram Silverman (ed.) *Man and Socialism in Cuba: The Great Debate.* New York: Atheneum.

Higman, B. W. (2002) 'The making of the sugar revolution'. In Alvin O. Thompson (ed.) *In the Shadow of the Plantation: Caribbean History and Legacy.* Kingston: Ian Randle, 40–73.

Højlund, Susanne (2014) 'Tasting time: the meaning of sugar in Cuba. Contextualizing the taste of sweetness'. In Vinicius de Carvalho, Susanne Højlund, Per Bendix Jeppesen and Karen-Margrethe Simonsen (eds) *Sugar and Modernity in Latin America: Interdisciplinary Perspectives.* Aarhus: Aarhus University Press, 73–89.

Howard, Philip A. (2015) *Black Labor, White Sugar: Caribbean Braceros and their Struggle for Power in the Cuban Sugar Industry.* Baton Rouge: Louisiana University Press.

Hulme, Peter (2012) 'Writing on the land: Cuba's literary geography'. *Transactions of the Institute of British Geographers NS* 37(3): 346–58.

Ibarra, Jorge (1986) 'Martí and socialism'. In Christopher Abel and Nissa Torrents (eds) *José Martí: Revolutionary Democrat.* London: The Athlone Press, 82–111.

Kapcia, Antoni (2000) *Cuba. Island of Dreams.* Oxford and New York: Berg.

Knight, Franklin W. (1990) *The Caribbean: The Genesis of a Fragmented Nationalism* (2nd edn). New York and Oxford: Oxford University Press.

Leitgeb, Friedrich, Sara Schneider and Christian R. Vogl (2015) 'Increasing food sovereignty with urban agriculture in Cuba'. *Agriculture and Human Values*: 1–12. http://link.springer.com/article/10.1007%2Fs10460-015-9616-9 (published online 18 June 2015).

Leogrande, William M. and Julie M. Thomas (2002) 'Cuba's quest for economic independence'. *Journal of Latin American Studies* 34: 325–63.

Lipton, M. (1977) *Why Poor People Stay Poor: A Study of Urban Bias in World Development*. London: Temple Smith.

McMichael, Philip (2006) 'Feeding the world: agriculture, development and ecology'. *Socialist Register*: 170–94.

Maingot, Anthony (1998) 'Global economics and local politics in Trinidad's divestment programme'. North-South Agenda Papers, no. 34, December. Miami, FL: Dante B. Fascell North-South Center.

Martí, Jose (1975) *Inside the Monster: Writings on the United States and American Imperialism* (ed. and transl. by Philip S. Foner). New York and London: The Monthly Review Press.

Massey, Doreen (2011[2005]) *For Space*. London: Sage Publications.

Meighoo, Kirk (2003) *Politics in a Half-Made Society: Trinidad and Tobago 1925–2001*. Kingston: Ian Randle Publishers.

Miller, Daniel (1994) *Modernity: An Ethnographic Approach. Dualism and Mass Consumption in Trinidad*. London: Bloomsbury Academic.

Mintz, Sidney (1974) *Caribbean Transformations*. New York: Columbia University Press.

Morales Domínguez, Esteban (2012) *Race in Cuba: Essays on the Revolution and Racial Inequality*. New York: Monthly Review Press.

Morris, Emily (2015) 'How will US-Cuban normalization affect economic policy in Cuba?' *American University-SSRC Implications of Normalization: Scholarly Perspectives on US-Cuban Relations*, April. https://www.american.edu/clals/upload/2015-AU-SSRC-Morris-How-will-US-Cuban-normalization-affect-economic-policy-in-Cuba-FINAL.pdf (accessed 8 February 2016).

Naipaul, V. S. (2001) *A Way in the World*. London: Vintage.

Olwig, K. F. (1985) *Cultural Adaptation and Resistance on St. John: Three Centuries of Afro-Caribbean Life*. Gainesville: University of Florida Press.

Pemberton, Rita (2002) 'The roots of survival: agriculture in Trinidad and Tobago during World War II'. In Alvin O. Thompson (ed.) *In the Shadow of the Plantation: Caribbean History and Legacy*. Kingston: Ian Randle Publishers, 405–21.

Pérez, Louis A. Jr (1986) 'Politics, peasants and people of color: the 1912 "race war" in Cuba reconsidered'. *The Hispanic American Historical Review* 66(3): 509–39.

——(1995[1988]) *Cuba: Between Reform and Revolution*. Oxford: Oxford University Press.

——(1998) *Cuba between Empires 1878–1902*. Pittsburgh, PA: The University of Pittsburgh Press.

Popkin B. (1998) 'The nutrition transition and its health implications in lower-income countries'. *Public Health Nutrition* 1(1): 5–21.

Portell Vilá, Herminio (1995[1939]) *Historia de Cuba*. Havana: Montero.

Potter, Robert B., David Barker, Dinnis Conway and Thomas Klak (2004) *The Contemporary Caribbean*. Harlow, UK: Pearson and Prentice Hall.

Premat, Adriana (2003) 'Small-scale urban agriculture in Havana and the reproduction of the "New Man" in contemporary Cuba'. *European Review of Latin American and Caribbean Studies* 75: 85–99.

Reardon, Thomas and Julio A. Berdegué (2002) 'The rapid rise of supermarkets in Latin America: challenges and opportunities for development'. *Development Policy Review* 20(4): 371–88.

Richardson, Ben and Pamela Richardson-Ngwenya (2013) 'Cut loose in the Caribbean: neoliberalism and the demise of the Commonwealth sugar trade'. *Bulletin of Latin American Research* 32(3): 1–16.

Richardson, Bonham C. (1992) *The Caribbean in the Wider World, 1492–1992*. Cambridge: Cambridge University Press.

Rodríguez-Ojea Menéndez, Arturo and Santa Jiménez Acosta (2005) 'Is obesity a health problem in Cuba?' *Human Ecology Special Issue* 13: 115–19.

Ryan, Selwyn (1973) *Race and Nationalism in Trinidad and Tobago*. Toronto: University of Toronto Press.

Sahlins, Marshall (1976) *Culture and Practical Reason*. Chicago and London: The University of Chicago Press.

Sanatan Dharma Maha Sabha (1989) 'Position paper of (SDMS) on the future of Caroni (1975) Ltd.' In *Future of the Caroni (1975) Ltd.: A Position Document*. 6 March, 1–10.

Sherwood, Marika (2003) *Origins of Pan-Africanism: Henry Sylvester Williams, Africa and the African Diaspora*. London and New York: Routledge.

Singh, Kelvin (2002) 'Race, class and ideology in postcolonial Trinidad, 1956–91'. In Alvin O. Thompson (ed.) *In the Shadow of the Plantation: Caribbean History and Legacy*. Kingston: Ian Randle Publishers, 444–63.

Smith, Joe and Jehlička, Petr (2013) 'Quiet sustainability: fertile lessons from Europe's productive gardeners'. *Journal of Rural Studies:* 148–57.

Thomas, Hugh (1971) *Cuba: The Pursuit of Freedom*. New York: Harper & Row.

Trouillot, Michel-Rolph (1992) 'The Caribbean region: an open frontier in anthropological theory'. *Annual Review of Anthropology* 21: 19–42.

Whitehead, Laurence and Bert Hoffman (2007) *Debating Cuban Exceptionalism*. London: Palgrave Macmillan.

Wilk, Richard (ed.) (2006) *Fast Food/Slow Food: The Cultural Economy of the Global Food System*. Lanham, MD and Plymouth: Altamira Press.

——(2013) '"Real Belizean food": building local identity in the transnational Caribbean'. In Carole Counihan and Penny Van Esterik (eds) *Food and Culture: A Reader* (3rd Edition). Oxford and New York: Routledge, 376–93.

Williams, Eric (2002 [1942]) *History of the People of Trinidad and Tobago*. Brooklyn, NY: A&B Publishers Group.

Wilson, Marisa (2010) 'The revolutionary revalorization of peasant production and implications for small-scale farming in present-day Cuba'. *History in Action* 1(1): 1–6.

——(2012) 'Moral economies of food in Cuba'. *Food, Culture and Society* 15(2): 277–91.

——(2013) 'From colonial dependency to "finger-licken" values: food, commoditization and identity in Trinidad'. In Hanna Garth (ed.) *Food and Identity in the Caribbean*. London: Bloomsbury, 107–19.

——(2014a) *Everyday Moral Economies: Food, Politics and Scale in Cuba*. Oxford: Wiley-Blackwell.

——(2014b) 'Agroecology and the Cuban nation'. In Yuson Jung, Jakob Klein and Melissa Caldwell (eds) *Ethical Eating in the Socialist and Postsocialist World*. Berkeley and New York: University of California Press, 167–87.

——(2016) 'Food and nutrition security policies in the Caribbean: challenging the corporate food regime?' *Geoforum* 73: 60–9.

Wilson, Marisa and Vishala Parmasad (2014) 'Political economies of sugar: views from a former sugar industry'. In Michael Goran, Luc Tappy and Kim-Ann Lê (eds) *Dietary Sugars and Health: From Biology to Policy*. Abingdon: Taylor & Francis, 13–26.

Wolf, Eric R. (1969) *Peasant Wars of the Twentieth Century*. New York: Harper Books.

World Bank (2008) *World Development Report 2008: Agriculture in Development*. Washington, DC. http://siteresources.worldbank.org/INTWDR2008/Resources/WDR_ 00_book.pdf (accessed 11 December 2015).

Afterword

Peter Jackson

This Afterword draws out some common themes and offers some reflections on the arguments made in this collection of essays on postcolonialism, indigeneity and the struggle for increased food sovereignty. The empirical range of the collection, based on ethnographic research in diverse parts of the world, provides an opportunity to take a comparative perspective on several of the key issues raised in the volume, of which I have chosen to highlight three. The first concerns the alterity of 'alternative' food movements where the empirical evidence presented here enables a more acute problematization of the rhetorical opposition of 'alternative' and 'mainstream' than has characterized many earlier conceptualizations. The second highlights the nature of (post)coloniality as a marker of persistence as well as of change, and the third concerns the interlinking of moral and political economies which are often cast in oppositional terms.

The alterity of 'alternative' food movements

One of the key provocations of the book is to ask whether a change of geographical focus challenges the principles that have defined 'alternative' food movements in other parts of the world. This transformative potential is explored through a shift in focus from the global North to the global South (with evidence from Africa, Latin America, the Caribbean, Asia and the Pacific), but also when attention is focused on more marginal places within the global North (such as Scotland in the United Kingdom or the coastal waters of British Columbia in Canada). Much of the literature on alternative food networks (AFNs) has been produced in the global North, based on examples from North America and Europe (including farmers' markets and community-supported agriculture, and movements such as Slow Food and *La Via Campesina*). The paradoxical qualities of these 'alternative' food movements have been widely noted, including their tendency to attract middle-class white consumers, able to pay the premium prices that such initiatives frequently involve (see, for example, Slocum 2007). The tendency towards co-optation or 'conventionalization' has also been examined, particularly in relation to the organic movement in California (Guthman 2004). While 'alternative' food movements derive much of their impetus from their oppositional

stance towards 'mainstream' market forces, opposing conventional or industrialized modes of production and the introduction of explicit ethical standards (as in
the case of Fairtrade), it is important to recognize that these distinctions are easily
overdrawn.

Wilson's Introduction reminds us that 'mainstream' food networks are as
geographically uneven and inhabited in multiple ways as 'alternative' food
networks, countering the assumption that local, small-scale production is inherently virtuous and that large-scale, globalized agri-food systems are inherently
exploitative (cf. DuPuis and Goodman 2005). Distinctions between alternative
and mainstream are in any case becoming increasingly blurred as 'alternative'
producers adopt the marketing practices of 'mainstream' food retailers and as
'mainstream' retailers appropriate the language of the 'alternative' sector. Even
the most industrialized foods are now being 'sold with a story' (Freidberg
2003: 4) about their provenance or other alleged qualities. Issues of provenance
are increasingly fraught as many foods now take a lengthy and complicated
journey 'from farm to fork'. The length and complexity of contemporary supply
chains was starkly revealed during the 2013 horsemeat incident when numerous
traders and subcontractors were found to be implicated in the adulteration of
beef supplies across Europe (Jackson 2015). But equally complex transnational
connections can also be found in the 'alternative' sector as Millner reveals in
Chapter 4 in her study of agro-ecological practices in Central America. In this
case, Millner shows how the discovery of permaculture in Australia by a
Salvadorian exile introduced these practices to El Salvador, aided by a community development worker from the UK who had encountered permaculture in
a Scottish eco-village. Millner argues that these imported discourses resonated
strongly with popular educational practices in Central America, but they arrived
there by a complicated route where they were adopted by permaculturalists who
drew on what Millner calls their 'mixed ancestry' rather than on a purely
indigenous heritage. Millner refers to a 'postcolonial ethos' whereby permacultural techniques and ideologies were re-appropriated in alignment with
indigenous histories and ontologies including the somewhat mystical idea of
Terra Madre (Mother Nature).

If the distinction between 'mainstream' and 'alternative' fails to capture the
complexity of postcolonial food politics, Ali and Vallianatos propose a more
varied vocabulary in Chapter 2 in their discussion of indigenous foodways in
the Chittagong Hill Tracts of Bangladesh. Building on the work of Fuller and
Jonas (2003),[1] the authors refer to the development of 'alternative-additional'
food networks where the Pahari people have shifted from forthright resistance to
partial adoption of 'mainstream' (for-profit) foodways, reproducing traditional
practices of food preparation and the values of sharing while incorporating
new agro-technologies and commercial relationships. A simple dichotomy of
resistance versus accommodation seems inadequate to the ethnographic complexity of the situation, including the Parahi's attitudes towards scientifically-
informed agricultural technologies, better described in terms of Homi Bhabha's
(1994) notion of (postcolonial) ambivalence.

(Post)colonial persistence and change

The editor and several of the contributors emphasize the importance of the brackets in '(post)colonial', highlighting the ongoing legacy of colonial structures of power and the unequal social relations that persist long after the formal ending of colonial rule. Despite what Byrd and Rothberg (2011: 4) refer to as 'the infamous and falsely periodizing "post" in postcolonial', it is clear that the effects of colonialism are far from over. The troubling prefix in 'post'-colonialism does not signify a decisive break with the past, as in a period that is over-and-done with, but the continuation of a vexed and persistent relationship whose 'power geometry' (Massey 1993) remains fundamentally asymmetrical. These arguments apply with equal force geographically as they do historically, for no part of the world has been completely untouched by the colonial project and, as Pat Noxolo and colleagues remind us, 'we are all part of a post-colonial world' (Noxolo *et al.* 2012: 425).

Accepting this logic, several contributors endorse Chakrabarty's (2009) invocation to 'provincialize Europe', acknowledging that Europe should take its place in the world, rather than providing the default position from which everything and everywhere else is regarded as a poor copy or inferior substitute. Applying these ideas to the field of consumption research, Miller advocates 'the equality of genuine relativism' whereby 'none of us is a model of real consumption and all of us are creative variants of social processes based around the possession and use of commodities' (1995: 144). Other models of comparative research, informed by a postcolonial sensibility, include Kuan-Hsing Chen's compelling assertion of the need to decolonize Western knowledge in *Asia as Method* (2010) where he insists upon using other Asian examples as points of comparison rather than always using British or North American case studies as exemplars of universal significance.

One of the present book's core arguments is to recognize the specificity of different colonial encounters and to trace the contemporary significance of diverse (post)colonial contexts. As Wilson outlines in her introductory comments on the varied configurations of alternative foodscapes, indigenous spaces should not be treated in the same way as other postcolonial spaces where indigeneity is not a key issue. An extant indigenous population marks out some contexts as distinctively different from others. As Wilson maintains, the concept of 'sovereignty' is itself a product of these circumstances, extending beyond autonomy and control over resources to the way these forces are shaped by the power of legal definition wielded by modern (sovereign) nation states.

These linguistic complexities belie simple notions of food sovereignty as the rights of people to define their own agricultural and food policies, advanced by groups such as *La Via Campesina* and discussed in more detail by Jarosz (2014). Indeed, in Chapter 5, Larder suggests that 'food sovereignty' is a deeply contested and highly gendered construct. Implicated with European concepts of property and ownership, food sovereignty can also be defined as the end of violence in the broadest sense, including violence against women. It involves access to farming knowledge and financial credit, agricultural machinery and artificial fertilizers, all of which are unequally available to men and women.

The distinction between indigenous and other postcolonial spaces also clearly arises in Chapter 3 in which Woodman and Menzies compare the contestation of fishing rights in Scotland, where a collective claim to the land has long since disappeared, with parts of British Columbia, where it is still directly relevant to indigenous people's political aspirations – a contrast which the authors suggest can be cast in terms of postcolonial versus anti-colonial movements. While the authors maintain that 'some of the people in the communities contesting fish farms in Scotland and Ireland might also be considered "indigenous"'(p. 75, this volume), the different social basis of these 'alternative' food movements must surely be acknowledged (as Wilson also notes in the Introduction).

Similar questions are raised by Morris and FitzHerbert in Chapter 1 on Māori food sovereignty in Aotearoa New Zealand where the 'indigenous imaginary' (that indigenous people are closer to nature than Western people) has a defining quality that is absent from other postcolonial situations. In the Māori case, an allegedly 'positive' discourse about the superior ecological wisdom associated with 'traditional' indigenous knowledge can be turned against those who espouse such knowledge (and those who are considered less 'authentic').

Notions of purity and tradition, deriving from Māori wisdom and unique ways of knowing, also arise in the cultural construction of Māori potatoes, which have provided a commercial opportunity for contemporary Māori to sell agricultural produce to a largely white (*pākehā*) clientele. But these notions are commercially limiting in a way that does not readily extend to other (non-indigenous) groups. When Māori appropriate 'Western' systems of authentication such as branding, food labelling and certification practices, the cultural politics of 'indigenous' food become even more complex. The commodity history of Māori potatoes is itself disputed and some varieties may have been introduced by early European explorers, reinforcing the need for scepticism about all origin myths and claims to culinary authenticity. A simple binary of mainstream and alternative seems woefully inadequate to capture the complexity of this particular form of 'cultural renaissance'. Some might argue that indigenous people are complicit in their own exploitation, 'selling out' their heritage or simply profiting from an available avenue – yet who would want to deny Māori growers this commercial opportunity? The fact that Māori are now facing competition from *pākehā* growers further accentuates Māori claims to cultural distinctiveness in what is becoming an increasingly crowded and entangled 'alternative' foodscape.

Similar issues over the representation of indigenous knowledge and the reinvention of tradition arise in Chapter 3, where Woodman and Menzies talk about First Nation fishing practices and the risks of ignoring 'traditional knowledge built up over centuries of living with the salmon' (p. 61, this volume), and in Millner's exploration in Chapter 4 of agro-ecological knowledge systems in El Salvador, where the author describes tradition as 'reimagined, not as a relic of a backwards past, but as a lively site of knowledge and expertise' (p. 82, this volume), drawing on '*cultural reservoirs* of knowledge and practices, shaped over many centuries' (p. 91, this volume, emphasis in original).

Moral and political economies

There are many examples throughout the book of the way that moral principles inform economic relations, belying the simplistic division between moral sentiments that are divorced from economic considerations and economic forces which operate through a brute calculus of profit and loss, unfettered by moral values or ethical concerns. Several chapters show how indigenous groups are cautiously adopting (and adapting) commercial farming and fishing practices, while 'mainstream' food producers in the West are appropriating some of the marketing devices from the 'alternative' sector. In Chapter 2 on indigenous foodways in Bangladesh, for example, Ali and Vallianatos refer to 'passion, tradition and identity', moral notions that play an important part in shaping Parahi food practices, while even the most commodified forms of food production are informed by ethical considerations, often articulated in terms of morally-charged notions of freshness, purity and nature (cf. Freidberg 2009).

In Chapter 3, Woodman and Menzies attempt to broaden the ethical landscape of 'alternative' foodways in their account of indigenous campaigns to oppose the extension of Norwegian fish farms in the waters off British Columbia. At risk of anthropomorphization, the campaigns are articulated in terms of 'justice for wild salmon'.[2] This might seem consistent with the relational ethics of Actor Network Theory (ANT), and its extension of ethical consideration to non-human actants, as outlined by Whatmore (1997) and others. But Woodman and Menzies are critical of some applications of ANT, accusing those who espouse such approaches of eliding questions of justice and sustainability (p. 59, this volume).[3] The chapter provides further evidence that there is no simple opposition between 'traditional' indigenous ways and 'modern' industrialized food production. In this case, some First Nations have accepted salmon farms as a way of generating employment, raising incomes and enhancing economic development, while others regard such encroachments as part of a wider process of cultural erosion, seen by some observers as amounting to genocide.

The links between moral and political economy are not just context-specific in a geographical and historical sense but can also be commodity-specific, taking a different shape in relation to different products, whether food staples or export crops. This is suggested by Millner in Chapter 4 on Salvadorian dependency on coffee cultivation and is taken further by Larder in Chapter 5 on irrigated rice in Niger, a crop whose commercial development was encouraged during the period of French colonial rule. It is also relevant to a younger generation of Caribbean consumers, some of whom prefer imported foods to locally grown 'slave foods' such as yams and other tubers (explored in Chapter 7 on Cuban exceptionalism). Different types of food and different modes of production receive differential evaluation according to the context, as McLennan explores in Chapter 6 on the 'failures' of community gardening on the Pacific island of Nauru where residents apparently prefer highly-processed, imported foods to locally grown fruits and vegetables. Welcomed as a positive development in the US on both health and environmental grounds, community gardens have not prospered in Nauru, despite high-profile interventions from the FAO's World Food Programme. While this could be taken as evidence of a moral failing on the part of the islanders, McLennan's work reveals the complex political history that has led to this dietary preference for imported fruit and vegetables, white flour and rice, refined sugar and tinned meat. These dietary choices are not merely a matter of taste preferences, social status or 'convenience' but are rooted in land ownership practices that raise difficult questions about how the produce of community gardens should be distributed in ways that are deemed fair and just. McLennan's ethnographic research shows how the fate of community gardens in Nauru is related to conflicting ideas of fairness, entitlement and value – moral notions that are linked to specific regimes of power. To blame Pacific Islanders for the consequences of their dietary habits (in terms of high levels of obesity and food-related non-communicable disease) would be to ignore this complex political history.

Moral and political economies are clearly intertwined in Woodman and Menzies' account in Chapter 3 of the unequal balance of power among those who support more intensive fishing of the global commons. The authors argue that policies that benefit major multinationals and corporate interests are being pursued by States and global actors such as the World Bank in the name of 'conservation' or 'development' – worthy causes that can be mobilized to support dubious commercial or political ends. The Cuban experience, explored by Wilson in Chapter 7, is also characterized by a highly moralized vocabulary, from José Martí's insistence on the value of hard work and self-sacrifice to the contemporary distinction between luxuries and needs. Despite the distinctive political and moral rhetoric, Wilson argues, the island's agricultural development followed a similar pattern to other Caribbean islands, such as Trinidad, in terms of the emphasis on export-oriented, industrialized production, based on a similar socio-technical paradigm, albeit one informed by a Soviet model rather than by a capitalist imperative. How Cuba's distinctive moral and political economy will be affected by the 'normalization' of relations with the United States, introduced by President Obama in 2014, remains to be seen although Wilson warns of the prospect of a return to Cuban dependency on US agri-food markets.

The chapters in this book clearly reveal the extent to which food security, food justice and food sovereignty are context-specific terms. One concept may have greater relevance in some situations, while another seems more appropriate in other contexts. One of Millner's Salvadorian informants articulates this idea with particular clarity, suggesting that food security in El Salvador is not just about access to good quality food but also about access to land (including issues of land ownership and governance). The term has very different connotations in other places, particularly in Europe and North America, where it is associated with other 'security' issues.[4] As part of a wider project on 'words with shadows', Itty Abraham (2009) has highlighted the distinctive connotations of *security* in English and how it differs from the use of its apparent homonym *segurança* in Brazil. In Chapter 4 Millner takes these translational issues further in the Salvadorian context, suggesting that what *cannot* be translated *should not* be translated, underlining the extent to which all knowledge is geographically and historically situated.

Given the complexity of the ethnographic contexts encountered throughout the book, it is understandable that authors are reluctant to frame their arguments about postcolonial and indigenous struggles in terms of the rather one-dimensional concept of 'resistance' with its implication of self-conscious, strategic and organized opposition to a clearly identified target. In the Introduction, Wilson proposes a different framing, referring to processes of 'subversion, contestation and appropriation' as an intrinsic part of the postcolonial histories and geographies of food production, exchange and consumption (p. 4, this volume). This is, perhaps, the key argument of the book, eschewing easy distinctions between local and global, alternative and mainstream, tradition and modernity, using ethnographically-grounded research to identify and map a more complex power-geometry in contemporary postcolonial and indigenous struggles for increased food justice and more sustainable practices of food production and consumption. The pursuit of greater historical depth and geographical specificity, in my view, leads to more subtle analysis, sharpening the book's political edge, increasing its contemporary significance and enhancing its global relevance.

Notes

1 Fuller and Jonas' (2003) typology of alternative economic spaces includes alternative-oppositional, alternative-additional and alternative-substitute spaces.
2 The authors later talk about 'letting the salmon speak' and of 'silenced fish'.
3 Woodman and Menzies go on to describe some accounts inspired by ANT as 'reductive', 'a-historical' and 'a-moral' (p. 74, this volume).
4 It might be suggested that concerns about food security, once largely restricted to the global South as a development discourse, only became a major concern in the global North following the spike in agricultural commodity prices in 2008. There was, of course, a wider context, which the UK's Government Chief Scientist Sir John Beddington described in terms of a 'perfect storm' involving climate change, population growth and resource depletion (Semple 2009).

References

Abraham, I. (2009) '*Segurança*/security in Brazil and the United States'. In C. Gluck and A. L. Tsing (eds) *Words in Motion*. Durham, NC: Duke University Press, 21–39.
Bhabha, H. (1994) *The Location of Culture*. London: Routledge.
Byrd, J. A. and M. Rothberg (2011) 'Between subalternity and indigeneity'. *Interventions* 13(2): 1–12.
Chakrabarty, D. (2009) *Provincializing Europe: Postcolonial Thought and Historical Difference*. Princeton, NJ: Princeton University Press.
Chen, Kuan-Hsing (2010) *Asia as Method: Toward Deimperialization*. Durham, NC: Duke University Press.
DuPuis, E. M. and D. Goodman (2005) 'Should we go "home" to eat? Toward a reflexive politics of localism'. *Journal of Rural Studies* 21(3): 359–71.
Freidberg, S. (2003) 'Not all sweetness and light: new cultural geographies of food'. *Social and Cultural Geography* 4(1): 3–6.
——(2009) *Fresh: A Perishable History*. Cambridge, MA: Harvard University Press.

Fuller, D. and A. Jonas (2003) 'Alternative financial spaces'. In A. Leyshon, R. Lee and C. Williams (eds) *Alternative Economic Spaces*. London: Sage, 55–73.

Guthman, J. (2004) *Agrarian Dreams: The Paradox of Organic Farming in California*. Berkeley: University of California Press.

Jackson, P. (2015) *Anxious Appetites: Food and Consumer Culture*. London: Bloomsbury.

Jarosz, L. (2014) 'Comparing food security and food sovereignty discourses'. *Dialogues in Human Geography* 4(2): 168–81.

Massey, D. (1993) 'Power-geometry and a progressive sense of place'. In J. Bird, B. Curtis, T. Putnam, G. Robertson and L. Tickner (eds) *Mapping the Futures: Local Cultures, Global Change*. London and New York: Routledge, 60–70.

Miller, D. (1995) 'Consumption and commodities'. *Annual Review of Anthropology* 24: 141–61.

Noxolo, P., P. Raghuram and C. Madge (2012) 'Unsettling responsibility: postcolonial interventions'. *Transactions of the Institute of British Geographers* 37(3): 418–29.

Semple, I. (2009) 'World faces "perfect storm" of problems by 2030, chief scientist to warn'. *The Guardian*, 18 March.

Slocum, R. (2007) 'Whiteness, space and alternative food practice'. *Geoforum* 38(3): 520–33.

Whatmore, S. J. (1997) 'Dissecting the autonomous self: hybrid cartographies for a relational ethics'. *Environment and Planning D: Society and Space* 15(1): 37–53.

Index

Page numbers in *italics* denote an illustration.